Coupling Constants of the Unified SuperStandard Theory
SECOND EDITION

We Find the Fine Structure Constant 1/137.0359801, and so
OUR UNIVERSE AND LIFE!
Also a Universal Eigenvalue Function for all Known Interactions,
And Running Coupling Constants to all Perturbative Orders

Stephen Blaha Ph. D.
Blaha Research

Pingree-Hill Publishing
MMXIX

ISBN: 978-1-7328245-4-6

Cover: The cover table depicts the fine structure constants of QED, and the Other Known Interactions. It also shows a plot of the QED eigenvalue function displaying the point where α appears.

Rev. 00/00/01 March 15, 2019

To My Grandchildren

Some Other Books by Stephen Blaha

All the Megaverse! Starships Exploring the Endless Universes of the Cosmos using the Baryonic Force (Blaha Research, Auburn, NH, 2014)

SuperCivilizations: Civilizations as Superorganisms (McMann-Fisher Publishing, Auburn, NH, 2010)

All the Universe! Faster Than Light Tachyon Quark Starships & Particle Accelerators with the LHC as a Prototype Starship Drive Scientific Edition (Pingree-Hill Publishing, Auburn, NH, 2011).

Cosmos Creation: The Unified SuperStandard Model, Volume 2, SECOND EDITION (Pingree Hill Publishing, Auburn, NH, 2018).

Immortal Eye: God Theory: Second Edition (Pingree Hill Publishing, Auburn, NH, 2018).

Unification of God Theory and Unified SuperStandard Model THIRD EDITION (Pingree Hill Publishing, Auburn, NH, 2018).

Calculation of: QED $\alpha = 1/137$*, and Other Coupling Constants of the Unified SuperStandard Theory* (Pingree Hill Publishing, Auburn, NH, 2019).

Available on Amazon.com, bn.com Amazon.co.uk and other international web sites as well as at better bookstores (through Ingram Distributors).

CONTENTS

FIGURES and TABLES

INTRODUCTION: SECOND EDITION

The hundred year pursuit of the value of the electromagnetic fine structure constant called alpha appears to have reached a conclusion in this book. In perhaps the happiest of theoretical circumstances it appears that alpha's value is determined by the internal dynamics of Quantum Electrodynamics (QED). Its origin is not cosmological or otherwise. In addition this book gives good reason to believe that the Weak and Strong interaction coupling constants are also determined internally dynamically. In all these cases no arbitrary external constants appear so there is no room for "fudging." The book also "explains" the difference in the values of Standard Model interactions. Running coupling constants are calculated and shown to have an asymptotic power law behavior with asymptotic freedom for non-abelian interactions, and increasing value for QED.

The study of these features is closely related to a universal form for the coupling constant eigenvalue functions. There is also a universal form for coupling constants of the form "alpha = v tan(h)" where alpha is a coupling constant, and v and h are numeric constants. This form is analogous to the Madhava-Leibniz expression for pi = arctan(1). Thus alpha is put on the same footing as the universal constant pi.

An important result of these studies is the implication that we are at the deepest level of fundamental elementary particle physics. There is no theoretical need for a deeper level and absolutely no experimental evidence either. Analogies with condensed matter physics suggesting a possible deeper level are interesting—but quite unlikely.

Perhaps the best support for the author's view that we have reached ground zero is the extraordinary accuracy of the theoretical QED predictions for hydrogen atomic levels, magnetic moments, Coulomb scattering, and so on. The agreement of theory and experiment is incredible making a yet deeper theory extremely unlikely. There are more phenomena likely in Physics but they will appear in new phenomena that augment known theory. Our books on the Unified SuperStandard Theory illustrate these possibilities.

Our theory of fine structure coupling constants supports our understanding of the features of the universe, the features of life, and the gamut of chemistry. This new understanding of the origin of coupling constants is a gusher!

The book is illustrated with many plots and diagrams as well as a summary of the Unified SuperStandard Theory. This edition contains the first edition with some minor clarifications. Most importantly, it uses an intermediate renormalization to map the QED eigenvalue function F_1 to F_2.

INTRODUCTION: FIRST EDITION

The coupling constants that appear in The Standard Model and The Unified SuperStandard Theory of the author have been the subject of much speculation. In this book we offer a procedure to calculate them in a relatively consistent manner using an extension of the Eigenvalue Function of QED. The application of this procedure, modified from the author's 1974 Physical Review D paper, leads to an accurate value for the Electrodynamics fine structure constant of $\alpha = 0.007297354$ compared to the known value $\alpha = 0.007297353$ (with inverses of 137.0359801 and 137.0359991 respectively.) Thus the value of α is intrinsic to QED and does not originate in cosmological or other Physical sources. The universe and Life, which are closely related to the value of α, are dependent on the internal features of QED.

In addition to determining α we also determine the ElectroWeak and the Strong SU(3) coupling constants to have values close to their known experimental values. We have a unique procedure based on Quantum Field Theory perturbative calculations that give approximately the known coupling constants.

The ability of our modified 1973-4 calculation of Eigenvalue Functions, together with the new insights into understanding the precise method to obtain coupling constant eigenvalues, is encouraging. It opens the possibility that The Unified SuperStandard Theory has within itself the mechanism for determining the constants appearing within it. It raises the hope that a similar self-determination mechanism may exist within the theory to determine the masses appearing in the Higgs particles sector of the theory.

INTRODUCTION

The coupling constants that appear in The Standard Model and The Unified SuperStandard Theory of the author have been the subject of much speculation. In this book we offer a procedure to calculate them in a relatively consistent manner using an extension of the Eigenvalue Function of QED. The application of this procedure, modified from the author's 1974 Physical Review D paper, leads to an accurate value for the Electrodynamics fine structure constant of $\alpha = 0.007297354$ compared to the known value $\alpha = 0.007297353$ (with inverses of 137.0359801 and 137.0359991 respectively.) Thus the value of α is intrinsic to QED and does not originate in cosmological or other Physical sources. The universe and Life, which are closely related to the value of α, are dependent on the internal features of QED.

In addition to determining α we also determine the ElectroWeak and the Strong SU(3) coupling constants to have values close to their known experimental values. We have a unique procedure based on Quantum Field Theory perturbative calculations that give approximately the known coupling constants.

The ability of our modified 1973-4 calculation of Eigenvalue Functions, together with the new insights into understanding the precise method to obtain coupling constant eigenvalues, is encouraging. It opens the possibility that The Unified SuperStandard Theory has within itself the mechanism for determining the constants appearing within it. It raises the hope that a similar self-determination mechanism may exist within the theory to determine the masses appearing in the Higgs particles sector of the theory.

The result would be a self-contained all-encompassing fundamental theory—the Holy Grail of fundamental Physics.

1. Some Highlights of the Unified SuperStandard Theory

This chapter lists some of the highlights of the Unified SuperStandard Theory. The next chapter describes the basis of the theory and lists its set of axioms. Readers familiar with the theory in Blaha (2018e) can skim or skip these chapters and proceed to chapter 3.

1. The number of spatial dimensions was determined to be the number of generators in the primary set of interactions of the space. In the case of an *empty* universe the primary set of interactions is the U(2) qubit transformations group. The number of U(2) generators is four and thus the dimension of space is 4 complex dimensions. Also and more importantly, considerations of Asynchronous Logic, and the requirement that physical processes must be able to proceed in parallel, require the number of spatial dimensions to be four. The book justifies four complex space-time dimensions with a Lorentz metric yielding Complex Lorentz group symmetry.

2. Boosts of the Complex Lorentz group transform a Dirac-like equation with a Landauer mass into four different forms (called species). Each form maps to a type of fermion: neutral leptons (neutrinos), charged leptons, up-type quarks, and down-type quarks. Neutral leptons and down-type quarks are tachyons. Some evidence exists for tachyonic neutrinos. Complex Lorentz boosts lead to the Complex Lorentz group factorization: SU(2)⊗U(1)⊗SU(3)⊗SU(2)⊗U(1). We map SU(2)⊗U(1)⊗SU(3) to fermion particle functional space to obtain the internal symmetry group for ElectroWeak and Strong Interactions: SU(2)⊗U(1)⊗SU(3). The remaining factors SU(2)⊗U(1) we map to the internal symmetry group for Dark Matter, which we take to be the Dark ElectroWeak Interaction (unconnected to normal matter interactions).

3. Parity Violation as seen in the Weak Interactions follows directly from the forms of the four types of fermions predicted by the Complex Lorentz Group.

4. The existence of four conserved (and partially conserved) quantum numbers such as baryon number and lepton number indicates that there is a U(4) group whose $\underline{4}$ representation causes each species to have four generations—three of the generations are known. We suggest that a fourth generation of much higher mass fermions exist.

5. In each generation there are four partially conserved quantum numbers. Thus we find that there is another U(4) group (called a Layer group) for each generation yielding the combined Layer groups $[U(4)]^4$. The $\underline{4}$ representation of each U(4) results in a fermion spectrum of four layers of four generations or 192 fermions in all. We see only one layer at present. The additional three layers of fermions remain to be found at much higher masses. The symmetry group of the Unified SuperStandard Theory is

$$[SU(2)\otimes U(1)\otimes SU(3)\otimes SU(2)\otimes U(1)\otimes U(4)\otimes U(4)]^4\otimes U(4)$$

where the last factor is for the broken Species group, which follows from Complex General Relativity.

6. Assuming all particles are massless at the Big Bang, and all particle types have an equal proportion of the total mass-energy then, we find that the 192 fermions and 192 vector bosons yield a Dark Matter percentage of 83.33% (experimentally the estimates are 84.5% and 81.5%). The proportion of Dark Mass-Energy is found to be 91% of the universe's mass-energy. Experimentally the proportion has been estimated to be 95%. These results agree well with experiment. See chapter 14 of Blaha (2018e) for details.

7. The instantaneous quantum effects between space-like separated parts of a quantum state ('spookiness') is taken to be a feature of fundamental importance. The only sensible way to implement this feature in quantum theory is to assume that the wave function of every particle is the inner product of a particle functional and a fourier

coordinate expansion. Particle functionals exist in a space with no distance measure. The space of coordinate fourier expansions also has no distance measure. Other functionals in a state (and their implicit coordinate fourier expansions) change *instantaneously* when one of the functionals comprising a state changes since coordinate space distance is irrelevant.

8. Fermion particle functionals are called *Qubes*. They exist 'within' every fermion. They have a mass that we take to be the Landauer mass—the minimal energy of a qubit. Boson particle functionals are called *Qubas*. They are assumed to be massless in the absence of all interactions to preserve free vector boson and spin 2 boson gauge symmetry. Free Higgs particles are assumed to be massless for consistency.

9. To have a completely finite theory with no infinities (including no fermion triangle infinities) we introduced Two-Tier Coordinates that replaced normal pointlike coordinates with a type of 'fuzzy' coordinates.

$$X^\mu = x^\mu + iY^\mu(x)/M_c^2.$$

10. Since the Unified SuperStandard Theory lagrangian would require higher order derivatives to account for quark confinement (linear potential terms) and for MOND-like deviations from conventional gravity, and since such terms would be outside a canonical lagrangian formulation, we introduced two fields for each particle (fermions and bosons) in a formulation we call Pseudoquantum Theory. Pseudoquantum theory enables a canonical lagrangian formulation. It has other advantages such as a clean separation of vacuum expectation values from quantum fields for Higgs particles. It also supports second quantization in arbitrary coordinate systems while maintaining the same particle interpretation of states in all coordinate systems.

11. The book also describes Higgs symmetry breaking and the use of the Faddeev-Popov Mechanism in detail for the theory.

12. Since a Complex Special Relativity requires a Complex General Relativity we considered Complex General Relativity and showed that it could be 'factored' into General Relativity and a new U(4) group that we called the Species group. Since Complex General Relativity must support interactions with all types of matter we specified a Species group interaction with all matter. Further, we assumed that the Species vector bosons acquired masses through the Higgs Mechanism The Higgs Mechanism caused Species group contributions to each fermion mass. Such a mass term would require each fermion particle mass to be both inertial *and* gravitational *solving the mystery of the equality of inertial and gravitational mass.*

13. We showed that the implicitly higher derivative Riemann-Christoffel curvature tensor for all interactions leads to new interactions beyond The Standard Model. In addition to yielding quark confinement and MOND-like modifications of gravity, it may help understand the missing nucleon spin issue, discrepancies in proton radius measurements, vector meson dominance (VDM), and so on.

14. We defined an Interaction Rotations group that caused rotations among all the vector boson interactions of The Unified SuperStandard Theory. We found that rotations that respected Superselection rules such as the Charge Superselection rule could have physical significance. One example is ElectroWeak Theory which is an application of Interaction Rotation transformations.

15. Since the number of fundamental fermions (192) and fundamental vector interaction bosons (192) is equal we considered Supersymmetric-like features of the Unified SuperStandard Theory.

16. The discovery of two new particles that do not appear to be within the framework of The Standard Model, as it is currently known, raises the possibility that they may be within the expanded fermion spectrum in The Unified SuperStandard Theory. Towards that end we present a *preliminary* assignment of the locations of the new fermions within the spectrum of the Unified SuperStandard Theory. Fig. 1.3

displays the “splitting” of the “Fermion Periodic Table” according to the set of interaction groups.

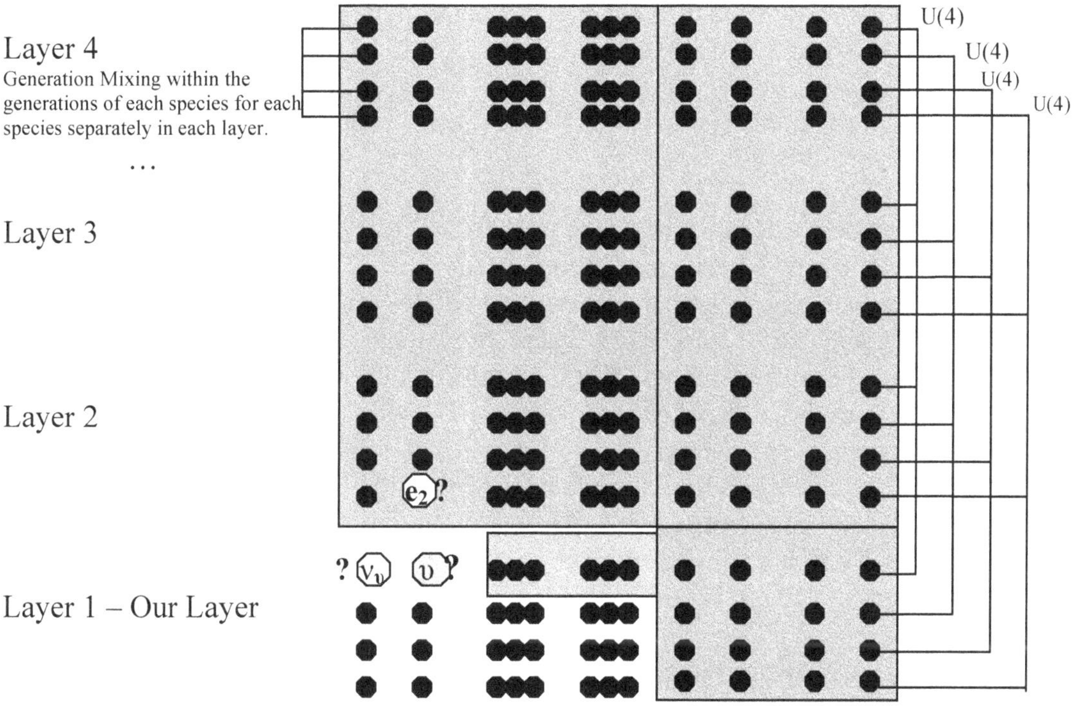

Figure 1.1. The Unified SuperStandard Theory fundamental fermion spectrum consisting of four layers of four generations of fermions. Dark parts of the periodic table are not yet found. Light parts are the known fermions..The circle labeled v_{υ} followed by a ? mark represents the possible new heavy neutrino where υ is a lower case Greek Ypsilon. If it is a fourth generation neutrino then we consistently name its fourth generation charged fermion “partner” as υ. The other circle labeled $\mathbf{e}_2$ followed by a ? mark represents a possible new heavy electron-like particle *of the second fermion layer.* This particle went through the earth without generating an interaction shower since it has a different electromagnetic interaction (due to a massive photon) from that of the first (known) layer. The lines on the left side exemplify the Generation group

mixing within one species. This Generation mixing applies to each species in each layer. The lines on the right side show the Layer mixing generation by generation.among all four layers for each generation individually. The left and right lines are partial examples of pattern of particle transformations of the Generation group and of the Layer groups..

Implicit in our discussion of the assignment locations of the newly found particles are the assumptions:

1. Fermion masses significantly increase from layer to layer. (Otherwise their particles would have been already found.)
2. Each layer has their own set of Standard Model interactions: Electromagnetic, Weak, Strong and their Dark Matter equivalents. The "electric" charge and internal quantum numbers of fermions are different from layer to layer. The interactions in the higher layers differ from those of the known layer by having larger vector meson masses.
3. We assume corresponding Standard Model interaction strengths (couplings) increase as one goes from layer to layer.
4. We assume Standard Model Interaction strengths are greater than the Generation group strength in each layer.
5. We assume all Layer groups interaction strengths are less than all the Standard Model interaction strengths and all the Generation group interaction strengths of all the layers. Otherwise there would be significant mixing between fermion layers.

The set of interactions (minus the Species group and the Interactions Rotation group of the Unified Super Standard Theory) are presented in Figs. 1.2 and 1.4.

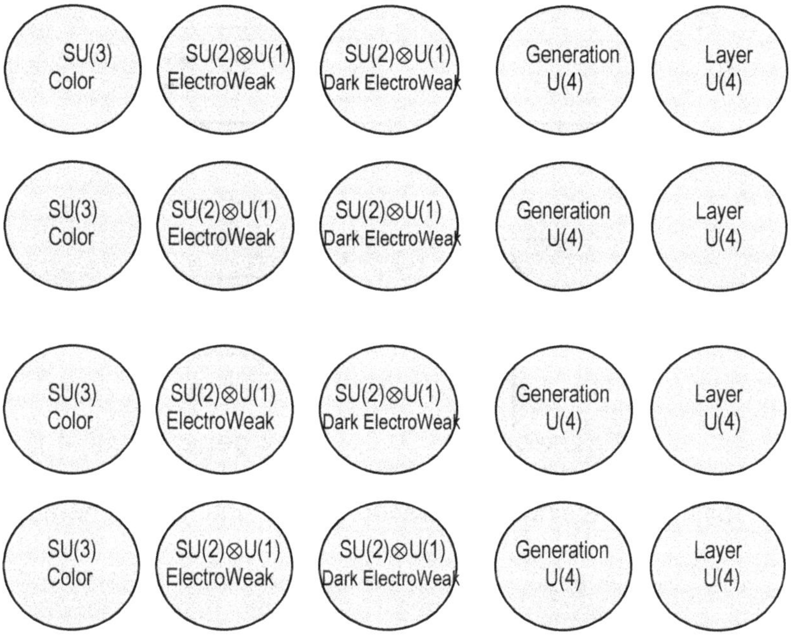

Figure 1.2. The set of four layers of internal symmetry groups corresponding to the four layers of spin ½ fermions and the four layers of vector bosons. In addition there are the Species group and the Interaction Rotations group Θ (not displayed).

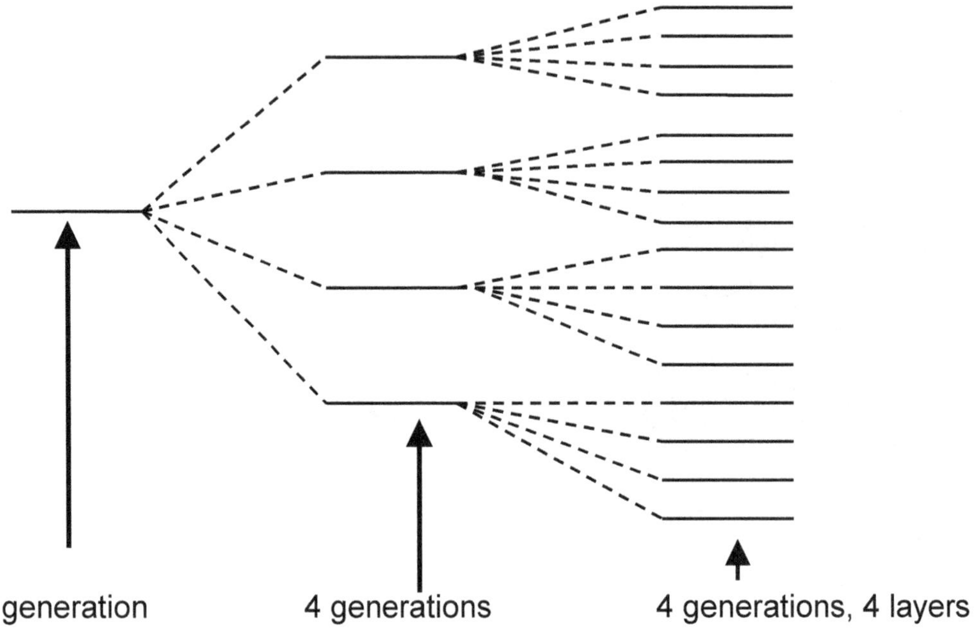

Figure 1.3. The "splitting" of a single generation fermion into four generations and then into four layers.

THE VECTOR BOSON PERIODIC TABLE

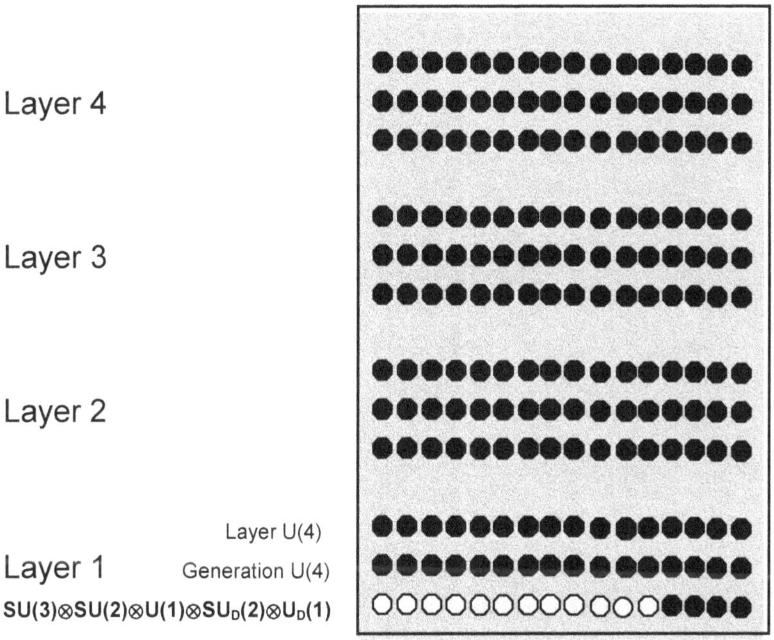

Figure 1.4. Each circle represents a group generator. The known vector bosons are in the lowest row with a white interior. Yet to be found vector bosons have a solid black interior. The Layer groups are distributed by layer symbolically although they each straddle all four layers.

2. Basics of the Unified SuperStandard Theory Derivation

In this chapter we will outline the process deriving our Unified SuperStandard Theory from fundamental axioms. Our theory has the somewhat unique feature of being derivable from first principles. Most other fundamental theories are constructed in an ad hoc fashion, often based on symmetry considerations, with features being added rather than derived.

The purpose of this chapter is to recapitulate the basis and derivation of the Unified SuperStandard Theory as presented in Blaha (2018e). It proceeds in the manner of Euclid with a clear connection between the steps of the derivation just as Euclid developed geometry from a progression of theorems.

In the third edition we presented a deeper foundation for the derivation of the Unified SuperStandard Theory that does not conflict with the derivation presented in earlier editions. It embodied a 'simpler' set of primitive terms and axioms, and 'explained' parts of the derivation in a cleaner way as well as providing *a new basis* for particle phenomena such as 'spooky' Quantum Entanglement.

Now we present the derivation in Blaha (2018e) for the sake of completeness and for the benefit of the interested reader.

We will begin with a brief historical view of the progress of fundamental physics from the cosmogonies of the Pre-Socratic Philosophers to modern views of Physics. Remarkably the conceptual thought processes of the Pre-Socratics is surprisingly similar to those of contemporary physicists: similar questions, similar proposals and similar qualitative views of reality – all based on limited empirical knowledge, limited mathematics, and limited means of proving assertions.

The great strength of the Pre-Socratics was perspicacity, a grasp of Logic and the ability to see beneath appearances.

2.1 The Historical Path from Pre-Socratic Cosmogony to Current Attempts at a "Theory of Everything"

Pre-Socratic efforts to understand physical reality and the Cosmos could be viewed as occurring in two phases: an initial phase where all happened as the result of the direct intervention of the Gods, and a second phase where physical materials and processes 'caused' natural phenomena and the invocation of a deity (deities) was not required.

2.1.1 Gods-Based Cosmogony

Prior to the 6th Century BC, Man's understanding of the Cosmos and physical phenomena were based on theological concepts – all events were based on the history and actions of the gods. In Mediterranean civilizations cosmology was well described by Hesiod's *Theogony,* and on Orphic and Zoroastrian cosmology that partly originated in Egypt, the Middle East and Persia. The origin of the universe was a history of the gods and their actions beginning from a primeval Chaos. Physical phenomena were the results of the actions of the gods.

2.1.2 Philosophic-Materialistic Cosmogonies

In the 6th Century BC a number of philosophers – notably Pherecydes – developed a materialistic view of the creation and Physics of the universe. Pherecydes postulated a primitive kronos that became the god Chronos – Time – from whom emanated a 'cosmic egg' and thence the gods Night and Eros. These gods created the constituents of the universe: earth, sky-air, and water. A reading of the historical fragments[1] describing this cosmology shows that much of the logic of their description of physical history and phenomena consisted of replacing theological terms with physical materialistic terms.[2] Thus there was a 'smooth' transition from theological descriptions to materialistic descriptions.

---

[1] See Kirk (1962) for a detailed study of Pre-Socratic Philosophy.

[2] This author pointed out this similarity of description in a paper (unpublished) in 1964. The similarity is reminiscent of the architectural similarity of wooden Greek temples which used pins to hold beams together and later stone Greek temples such as those of the Parthenon which placed decorative marble 'pegs' in the temple in imitation of the earlier wooden temples' pegs.

Various philosophers developed 'Theories of Everything' from this point on. The best known theory is the four element theory: fire, wind, water and earth of the 5th Century BC.

Miletus was a hotbed of Philosophic thought with Thales (and Homer) in the 6th Century BC holding that water was the sole element from which all elements were generated. He was followed by Anaximander who posited a sophisticated concept – 'the Indefinite Apeiron' – a single primary element from which the four elements were generated as well as intermediate forms of these elements. Anaximander also proposed 'plural worlds of infinite extent' – a concept familiar in various 'modern' forms since the 16th Century including our Megaverse proposal.

Anaximenes of Miletus suggested that the apeiron was air. The philosopher Diogenes also postulated that all was ultimately composed of air.

A succession of philosophers made noteworthy suggestions:

1. Theophrastus proposed that all matter was in eternal motion – dynamics.
2. Heraclitus proposed the unity of all things with fire as the primary element.
3. Democritus and Leucippus proposed matter was made of atoms of an infinite variety of shapes (since there was no reason why they should all have the same shape).
4. Xanophanes suggested that all things were made of Earth and water.
5. Pythagoras suggested all was based on Number as an extension of apeiron to unity with all other numeric quantities (multiplicity) generated from it.

Thus the concepts of materialistic Physics have a clear beginning in the Pre-Socratics. Notably, philosophers were quick to seize on the notion of a primary form of matter from which the varieties seen in Nature were derived. The Indefinite Apeiron of Anaximander is an especially noteworthy conceptual precursor to our current theories of elementary particles.

A reading of the fragments of the Pre-Socratic philosophers writings raises a number of significant questions that do not appear to have been seriously considered:

1. Why and how did the transition from a theocratic description of the universe to a materialistic description take place? It is thought that Pherecydes was the first (or the leading person) in proposing this change of *zeitgeist*. But the source of the concept is elusive since the observation of Nature was rudimentary and the leap of thought – contrary as it was to prevailing religious thinking – is substantial – inspired!
2. After the transition to a materialistic/scientific thinking pattern, how did the observed multifold aspects of Nature become viewed as based on a appearances of a few substances (fire, air, water, and earth or one or more of these substances)? We can see that transformations between substances such as water upon heating becoming 'air' but to have a 'Theory of Everything' based on a few, or one, fundamental substance is a leap of thought reminiscent of the 20th century attempts at a Theory of Everything such as String theory.

No ready answer to these questions can be found – mostly due to the loss of books when the Great Library of Alexandria was burned. The burning of the Library and the destruction of the first group of 'scientists' engaged in the study of natural phenomena[3] (together with the murder of Archimedes in Syracuse) shows that war can have a major retrograde effect on civilization.[4]

If one must have an answer one can only suppose that these advances – and they were major conceptual advances – were the result of 'inspired' creative thought. A possible view of this form of creativity is perhaps embodied in the story of Newton and a falling apple. The natural view is to see the fall of a physical apple. However, a mind can reconstruct this view to see an instance of a force of nature – gravity. What one sees, can be interpreted differently in the 'mind's eye.' And lead to a new, different perspective.

---

[3] Such as a prototype steam engine.

[4] Of course today's world is different with major science and engineering efforts for war. Today we face the issues of the obliteration of civilizations.

2.1.3 Atomic Theory

From the time of the Pre-Socratics to the mid-19th Century the nature of matter was a subject of dispute with some arguing matter was continuous and others arguing that matter was ultimately composed of atoms. The dispute was poignantly brought to a head by the success of the atomic theory of Boltzmann and others. Their success was controversial and vigorously condemned by opponents for decades.

By 1900 the atomic theory of matter was largely accepted and the door was opened to the successes of quantum theory.

2.1.4 Twentieth Century Physics

With the acceptance of the atomic theory of matter a series of developments principally related to atomic spectra and black body radiation led to the development of quantum mechanics in the 1920's, and quantum field theory in the 1930's and 1940's. The great experiments from the 1930's through the 1970's led to the Standard Model of elementary particles and their interactions.

Starting from the nineteen-teens a series of efforts were made to create a unified theory of everything with noteworthy initial efforts by Weyl, Einstein and Eddington. The efforts took on a new life with the development of unified quantum field theories and of string theories in the 1970's and afterwards.

In an effort to avoid the *ad hoc* nature of these theories which assumed group symmetries and/or strings, this author developed a theory derived from fundamental principles in a manner reminiscent of Euclidean geometry that culminates in this book since it describes the known properties of matter, proposes new possible features, and opens the door to extensions to new phenomena should they be discovered in future experiments.

This chapter now begins the derivation by specifying the requirements of the theory, defining primitive terms and then stating the theory's basic axioms à la Euclid. There are two advantages of a derivation from sound axioms: 1) the resulting derivation shows what must be included in the complete theory; and 2) the derivation automatically excludes possibilities that a creative theorist might think to include in the theory.

2.2 Underlying Physical Principles of Creation

2.2.1 Creation by a Cosmic Pseudo-Euclid Entity

The derivation of the Unified SuperStandard Theory seems to be best presented within the framework of a putative being who may or may not be a 'real' entity. Of this 'being', which might have been called the Cosmic Pseudo-Euclid Entity, but we will call the Entity, we can say: 1) It must be 'outside of' (independent of) time, 2) It must be 'immaterial' (not composed of anything), 3) It must be 'unchanging' in itself and 4) It originates the derived theory presented here.

In earlier editions we have called this Entity the 'Unmoved Mover' that implements the theory as Reality and causes all events to happen in a manner consistent with the Unified SuperStandard Theory.[5]

In this section we start with a discussion of the basic prerequisites of Creation (a general theory of everything) and then proceed to specify the axioms of creation that yield the derivation.

2.2.2 Creation Dynamics Rationale

The derivation of the theory for our universe (and for other possible universes[6] within a Megaverse) has certain fundamental prerequisites if one wishes to have dynamical physical processes, as we know them, to occur within interesting universes. We take these prerequisites for granted normally and proceed to concoct theories of physics as if they may be pulled out of a hat like the proverbial rabbit. However any attempt to create a universal physical theory must meet these prerequsites to build a theory from fundamental primitive terms and axioms in a manner similar to Euclid's construction of geometry.

---

[5] The role and features of the Entity would suggest that it is God to many and is discussed earlier in the God Theory section. However, since we address only Physics issues, we shall leave its nature an open issue in the Unified SuperStandard Theory derivation.

[6] We will call other universes *exoverses*.

2.2.3 Fundamental Prerequisites for a Fundamental Theory of Physics

We can list fundamental prerequisites based on a general knowledge of the necessary nature of a fundamental theory of Physics. This approach presumes a general knowledge of the theory that we wish to construct illustrating the maxim, "Our ends determine our beginnings."

A. A time variable must exist that may have various forms.

B. We wish to have a dynamical fundamental theory that evolves in time. Thus there must be a mechanism(s) that allow dynamical processes to exist that may, or may not, run in parallel.

C. Multiple parallel processes can execute.

D. There must be a space with a coordinate system(s), and distance measure, within which processes can execute.

E. There must be particles upon which dynamical processes execute.

F. There must be a space of functionals that support the creation of particle states and help determine their properties. The particle functional space frees particles from a complete dependence on coordinate space.

G. There must be a space of 'waves' of free field fourier expansions for all the fundamental particles absent interactions..

H. There must be an order in the 'created' dynamical theory that is embodied in a form of a computational language[7] with a Chomsky-like *Grammar* using a

[7] The possibility that the universe is one enormous Word was explored in *Cosmos and Consciousness* (Blaha (2003)) in physical, philosophical, and religious contexts. A few years ago around 2012 the author found a book with a similar title by R. M. Bucke published in 1901 entitled *Cosmic Consciousness* on the evolution of Man to a new level

finite set of terminal and nonterminal *symbols* that constitutes an alphabet (vocabulary).[8] The ordering in the form of a language with grammar *Production Rules* ensures the consistency of the generated theory.[9]

I. Creation should opt for Vitamorphic[10] universes that support life in some form. Recent studies have shown that evolution favors the development of increasingly intelligent life. Thus the ultimate appearance of intelligent life at places within universes appears to be natural—making the *Anthropic Principle* an evolutionary consequence[11] of the *Vitamorphic Principle.*

These prerequisites would seem to be necessary and sufficient for the specification of primitive terms and axioms for a fundamental theory.

2.3 Basis of SuperStandard Axioms

In the following sections we present a revised set of 'primitive' terms and axioms for our theory. A comparison of this new set of axioms with those provided in earlier editions will show that they are equivalent except for a few new axioms. They are also more simply stated, have fewer overlaps between axioms, and cleanly lead to our theory of elementary particles.

---

of consciousness. The content of this book is unrelated to Blaha (1998) – first edition - and (2003) as well as Blaha's other books.

[8] Particle Computer Languages are described in Blaha (2005b) and (2005c) as well later in this volume and in other books by the author.

[9] See chapter 8 of Blaha (2018e) for definitions and details.

[10] The *Vitamorphic Principle* states that universes should support some form of life realizing that there are many varieties of life and borderline forms of life. A 'tight' definition of life has not been satisfactorily constructed. There are many borderline entities that may or may not be called life. We take 'Vitamorphic' to mean 'life enabling' in English. Vitamorphism is not a concept without meaning—a universe (Megaverse) consisting of only inert matter without energy present would be non-Vitamorphic. The Anthropic Principle, briefly put, states that intelligent human-like life should exist.

[11] One can well wonder whether the emergence and dominance of Mankind has eliminated the possibility of the emergence of other intelligent species on earth from the many semi-intelligent species that exist now and in the past.

The goal of this edition is to derive the Unified SuperStandard Theory in the manner of Euclid with a clear connection between the steps of the derivation just as Euclid developed geometry from a progression of theorems.

2.4 Primitive Terms and Axioms

Primitive terms can be as simple as those of Euclid or they can be more complex. The level of simplicity depends on the nature of the theory and the Physical Laws that emerge from it. In the case at hand, a fundamental unified theory, the constructs that emerge in the construction of the theory are mathematically complex. Consequently, the choice of primitive terms and axioms may be expected to be mathematically complex as well, unless one wishes to expand the primitive terms into a more detailed, term by term description in simpler, more basic primitives. We will not pursue that alternative here since the terms that we use are 'self-explanatory' to the Elementary Particle Physics theorist knowledgeable about quantum field theory and particle symmetries.

2.5 Mathematics and Conceptual Prerequisites

Due to the complexity of the Theory we have chosen to specify mathematics prerequisites and use them in the derivation rather than devoting parts of the derivation to mathematical preliminaries. Therefore we use complex variable theory, Riemannian coordinates, group theory, classical and Quantum Logic, functionals, Chomsky-like computational languages, and so on without bringing in unnecessary supporting details from them.

We also assume certain physical concepts such as distance, quantum features, second quantization, covariance under a group transformation, and spatial curvature.

The axioms use some of these prerequisite concepts treating them as primitive terms for the derivation.

2.6 Primitive Terms for the Unified SuperStandard Theory

The set of primitive terms of the theory are:

Qubits
Qubes
Qubas
Core
Grammar
Terminal and Nonterminal Symbols
Production Rules
Speed of Light
Spatial Dimensions
Space and Time Coordinates
Covariance under group transformations
Asynchronous processes
Parallel Processes
Reference Frame
Complex Lorentz Group
General Coordinate Transformations
Gravity
Universe
Particle Masses
Fermions
Bosons
Particle States
Particle Rest State
Particle Momenta
Spin
Canonical Quantization
Quantum Process
Quantum Entanglement
Second Quantization
Quantum Field Theory
Quantum States
Asymptotic Particle States
Internal Symmetries
Coupling Constants
Discrete Symmetries
Yang-Mills Local Gauge Theory
Functionals
Functional space

In choosing these primitives, we understand that they generally embody a significant theoretic description or body of knowledge. We do not include names used in the mapping to reality (such as quark) in the list of primitives since the mapping to reality is a separate issue in our view.

2.7 Axioms for the Unified SuperStandard Theory

The set of axioms that we list below is supplemented by the Decision Axioms of Appendix C.1.3 of Blaha (2018e). The 'new' physical axioms are

PARTICLE AXIOMS

1. All matter and energy is composed of particles.
2. Each fundamental particle has a physico-logic structure within it that we designate its core.
3. Particles form an alphabet with a finite number of characters and combine in ways specified by the quantum probabilistic production rules of a quantum computational grammar.[12]
4. A core is a particle functional that combines with a free field fourier coordinate expansion in an inner product to produce a free second quantized particle field.
5. There is a 4-dimensional space of particle functionals, called *particle functional space*, with the distance measure, eq. 2.1 below, specifying the transformation group of particle functionals.
6. Particle functional space consists of a single point.
7. The core of a fermion functional is called a *qube*. Fundamental bosons have a core consisting of a boson functional called a *quba*.
8. Qubes have a a bare mass. Qubas have zero mass.

SPACE AXIOMS

9. The dimensions of a coordinate space-time are determined by the number of fundamental[13] interactions, and the requirement that all parallel processes, with parts perhaps separated by distances, can occur synchronously.

---

[12] See Blaha (2005b).

10. Spatial coordinates are inherently complex-valued.
11. Space has one complex-valued component that plays the role of time. Physical phenomena dynamically evolve based on the time variable.
12. The infinitesimal distance ds between two space-time points is given by

$$ds^2 = dt^2 - d\mathbf{x}^2 \tag{2.1}$$

where d**x** is a vector of the spatial coordinates. Transformations between coordinate systems preserve the value of ds and define a transformation group. (The Complex Lorentz Group)
13. Physically acceptable reference frames have real-valued coordinates. These coordinates can be obtained by group transformations from complex-valued coordinate systems. Physical space-time measurements are made in a real-valued coordinate system.
14. The speed of light is the same in all reference frames.
15. Free fundamental leptons must have a real-valued energy.
16. Gravity may cause space-time to be curved. (Complex General Coordinate transformations[14])

DYNAMICS AXIOMS

17. The complete theory has a lagrangian formulation. If the lagrangian is truncated to quadratic form (interactions set to zero) then symmetries appear that are the source of particle symmetry groups that persist with broken symmetry after interactions are reintroduced. The lagrangian specifies a set of production rules

---

[13] Interactions that would exist in the absence of fermion particles.

[14] If the metric tensor of space-time is analogous to one of the metric tensors of the superfluid phases of ^{3}He, then space-time might have several metric tensors in 'various regions.' If the space-time metric tensor is analogous to the ^{3}He-B superfluid phase metric tensor, which has an effective gravity with a complex metric tensor, the space-time metric tensor would be the familiar one of General Relativity. However if the space-time metric tensor is analogous to the metric tensor of superfluid ^{3}He-A, which exists at higher pressure and temperature, then the space-time metric tensor might be similar to the Penrose twistor theory metric tensor. In this case the corresponding General Relativity may have a twistor-like metric tensor: perhaps in the early universe, and/or inside black holes, and/or in small universes with higher pressure and temperature than our universe. We will assume the conventional metric for Complex Special and General Relativity.

of a type 0 Chomsky language generalized to include production rules for the generation of all strings of symbols (particles) from any strings of symbols (including the *head symbol.*)[15]

18. The lagrangian of the theory must be invariant under coordinate system transformations.
19. Dynamical particle equations must be covariant under group transformations.
20. All interactions have a local Yang-Mills gauge theory formulation.
21. The vector bosons, and the interactions among them, are determined by terms in complete lagrangian, some of whose parts are obtained from the Riemann-Christoffel Curvature Tensor.

QUANTIZATION AXIOMS

22. All fields must be canonically quantized.
23. Fermion and Boson vacua can be defined that are valid in all coordinate systems.
24. The number of particles in an asymptotic state of any given type is invariant in all reference frames.
25. Quantum processes starting in an initial quantum state, with parts separated by a distance after a time, can have the parts synchronously change each other instantaneously. (Quantum Entanglement)

2.8 The Derivation of the Unified SuperStandard Theory

The derivation of the Unified SuperStandard Theory has been a multi-year process undertaken by the author. Much of the derivation appears in Blaha (2015a), (2016f), (2017b), (2017c), (2017d), (2018a), (2018b), and (2018c). Earlier work, upon which these books are based, is referenced in these books and listed in the References in this book.

We now show a clear logical development of the Unified SuperStandard Theory from first principles in a manner reminiscent of the derivation of Euclidean geometry. This derivation will be seen to be primarily based on a 'simple' concept—the space-

---

[15] Chapter 8 of Blaha (2018b) discusses computational languages for particles in detail.

time in our universe. The derivation explains the construction process of the physical theory. The manner of the derivation embodies the mapping of the theory to physical reality. The question of the 'Unmoved Mover' (Appendix C of Blaha (2018e)) is necessarily beyond the scope of Physics. (See the God Theory section of Blaha (2018e).)

2.9 The 'Time' Order of the Derivation

An examination of the derivation presented here will show that the derivation has an implicit ordering which is analogous to the ordering of theorems in Euclidean Geometry. It begins with the most 'primitive' aspects of the theory and progressively develops and adds features.

Thus it is not time-ordered like the dynamical progression of the universe from the Big Bang, or like the dynamical evolution of physical processes in time. To some extent the derivation consists of a derivation of the basic nature of the fundamental fermion and boson spectrum, and the Standard Model interactions, from a Complex Lorentz group analogy; a derivation of the number of fermion generations, and the Generation group interaction, from the form of the free lagrangian fermion terms and their associated conservation laws; a derivation of the fermion layers, and the Layer groups interactions, from conservation laws implicit in the free lagrangian plus ElectroWeak terms; and a derivation of boson layers from Complex Lorentz group considerations.

2.10 Emergence vs. Derivation

Emergence is an interesting approach that is being studied in Physics and Biology. The use of emergent concepts to 'derive' the features of complex phenomena from simpler constructs is conceptually different from a derivation along Euclidean lines. However the procedure is quite similar. The most significant difference is that emergence suggests that complex phenomena somehow mask an underlying simpler *dynamical* structure at their base. The dynamics of the basic structure then generate complex phenomena. The author believes the emergent approach introduces a deeper layer of Physics for which no experimental evidence exists.

An examination of the derivation of the Unified SuperStandard Theory presented here shows that the primitives of the derivation – terms and axioms – may be sufficiently 'masked' so as to justify describing our derived model as quasi-emergent, since it is based primarily on the Complex Lorentz group and Complex General Relativity.

3. Missing Additional Features of the Unified SuperStandard Theory

The Unified SuperStandard Theory derives its form from a set of axioms. It does not determine the coupling constants or the particle masses of the theory. In this book we calculate the coupling constants of the Electromagnetic, Weak and Strong interactions of the first layer of fermions. The other fermion layers may have similar coupling constants for their equivalents to these interactions or they may differ. Also the interactions strengths in all layers are affected by their accompanying vector boson masses. This is particularly evident currently for the Weak interaction vector bosons.

3.1 Determination of Coupling Constants

The renormalized coupling constants of The Unified SuperStandard Theory can be determined experimentally. However their theoretical calculation is uncertain. This gap in our understanding suggests that there is a major aspect of fundamental physics that is not understood. The fact that we can determine the form – but not the values of coupling constants – so directly from basic principles suggests that a new basic principle(s) is needed to complete The Unified SuperStandard Theory. A similar comment applies to fermion and boson masses – both our mechanism,[16] and the vanilla Higgs Mechanism, arbitrarily fixes particle masses. (Attempts to relate particle masses using various symmetries beg the question. As Isidore Rabi (Columbia) once said in a different context, "Who ordered them?" Proposed symmetries are typically "pulled out of a hat.")

The *most* meaningful attempt to determine a coupling constant in a non-trivial 4-dimensional quantum field theory was that of Johnson, Baker and Willey[17] in a 4-dimensional model – massless Quantum Electrodynamics. They developed the theory to

---

[16] Blaha (2015c).
[17] M. Baker and K. Johnson, Phys. Rev. **D8**, 1110 (1973) and references therein. See Appendix C.

the point where if one function, that they called the eigenvalue function, had a zero at the value of the fine structure constant[18] α = 1/137.035999139 (31) at $Q^2 = 0$ then the theory would have no infinities.[19]

This author then developed an approximate solution for the eigenvalue function in perhaps the most comprehensive 4-dimensional quantum field theory calculation to all orders in α. The approximate calculation agreed with known exact results to 6th order in e. In 1974 this author[20] did not find an eigenvalue function zero at the known value of the renormalized α for reasons that will be elucidated later. The remedy for this deficiency will also be described later.

The Johnson, Baker, Willey model QED eigenvalue condition illustrates one possible approach to determining the coupling constants of QED and possibly the other coupling constants of The Unified SuperStandard Theory. It appeared possible that eigenvalue conditions might fix coupling constant values. In this book we will suggest that the non-aabelian group interactions of the Unified SuperStandard Theory may possess eigenvalues conditions similar to that of massless QED. We will then calculate approximate values of coupling constants.

What other approaches are possible? There is an anthropomorphic approach which posits the necessity of certain ranges of some coupling constants for human life, and life in general, to exist. We are not comfortable with this approach since it seems to "beg the question." The input is equivalent to the output mitigating its character as fundamental.

One could also study the set of coupling constants in a 10-dimensional (or other dimensional) space looking for the set of coupling constant values.

These alternate possibilities are not viable at present. So we will proceed with the eigenvalue function approach, fixing the apparent failure in Massless QED, and developing eigenvalue functions for the coupling constants of other interactions.

---

[18] C. Patrignani et al, (Particle Data Group), Chin. Phys. C**40** 100001 (2016).

[19] An alternate summation in perturbation theory gives the zero at the bare coupling constant α_0. Both alternatives will be discussed later where the renormalized coupling constant will be shown to be the correct zero.

[20] Equation 1 in our paper S. Blaha, Phys. Rev. **D9**, 2246 (1974).

3.2 Origin of Particle Masses

Particle masses were fixed by either the original Higgs Mechanism or by our new mechanism that was based on an extension of Quantum Field Theory to include classical fields, that contain vacuum expectation values, that cropped up in the original Higgs Mechanism and were handled "by hand." (See Blaha (2015c).)

The origin of the constants appearing in either approach to particle masses is unknown at present. They are *ad hoc* parameters inserted by hand. This book will not investigate their origin.

3.3 Principle of Unfolding Depth

There are many situations in Physics where an initial "discovery" leads to deeper and deeper levels of understanding. The clearest example of this process of unfolding depth is the sequence of discoveries of the nature of matter:

1. Four basic elements: earth, air, fire, water
2. The chemical elements of the eighteenth and nineteenth centuries
3. The discovery that the elements are made of atoms by Boltzmann and others in the late nineteenth century
4. The discovery that atoms consist of electrons circling a nucleus in the early twentieth century
5. The discovery that a nucleus is composed of protons and neutrons
6. The discovery that neutrons and protons are composed of quarks

This example of unfolding depth can be multiplied many times in Physics and Chemistry. The case most relevant to the current discussion is that of Electromagnetism where we see:

1. Classical Electromagnetism
2. Renormalizable Quantum Electrodynamics
3. Incorporation of Quantum Electrodynamics within the Weinberg-Salam Model with a complex renormalization program (due to t'Hooft and others).

4. Possible incorporation of the Weinberg-Salam Model within a unified theory such as our Unified SuperStandard Theory with the elimination of renormalization infinities using Two-Tier coordinates.

The above Electromagnetism sequence has the property that the theory in the earlier stages is still correct within its range of validity. So we can still perform computations in Quantum Electrodynamics using Pauli-Villars regularization (stage 2 above) and the Gell-Mann-Low QED renormalization studies are still valid.

Having seen how Physics theories unfold to greater depth in the above examples (which could have been readily expanded to a much larger set of examples), we now propose a simple principle of Unfolding Depth.

Principle: Theories of Physical Phenomena tend to unfold to greater depth with the passage of time in such a way that earlier stages of the unfolding process still retain their validity to a great degree.

4. Massless Quantum Electrodynamics

Quantum Electrodynamics (QED) has been presented in depth in many papers and textbooks. In its pristine form it contains a massive electron interacting with a massless photon. Due to these interactions the electron mass and the electric charge are renormalized in perturbation theory. The renormalizations lead to infinities that have been of some concern in the early days of QED. Starting with Pauli-Villars regularization, techniques have been developed to control these infinities so the theory makes only finite, measurable experimental predictions. The theoretical predictions of QED have achieved a remarkable degree of consistency with experiment to make QED the most accurate of all theories. Successes include the hydrogen atom, particle magnetic moments, the Lamb shift, and short distance Coulomb scattering – all calculated in QED with up to ten digit accuracy!

4.1 The Importance of the Value of α

Perhaps the most important question in QED is the origin of the value of the QED coupling constant, the Fine Structure Constant denoted α, with the measured value[21] of $\alpha = 1/137.035999139\ (31)$ at $Q^2 = 0$. *Many features of Nature and Life have been shown to depend significantly on the value of α.* The form of the universe is largely determined by the value of α and gravity. Chemistry depends to great detail on the value of α. Life is based on Chemistry, as noted by Paracelsus in the 16th Century.

The prodigious importance of the value of α has led to many attempts to determine (or calculate) its origin: attempts ranging from the subatomic to the cosmological. All these efforts have been unsuccessful. Perhaps one of the better known efforts was that of W. Heisenberg who determined α to have a value of about 3. Other efforts, including a past effort of this author, have also been unsuccessful.

---

[21] C. Patrignani et al, (Particle Data Group), Chin. Phys. C**40** 100001 (2016).

In this book we accurately determine the approximate value of α. We shall show that the value of α is determined by an eigenvalue function whose zero removes massless QED divergences. This zero occurs at the measured value of α.

4.2 The QED Divergences Requiring Renormalization

QED has divergences when calculated in perturbation theory. The divergences can be isolated into four divergent quantities:

Z_1	- vertex renormalization factor
$Z_2 = Z_1$	- self-energy renormalization factor
Z_3	- vacuum polarization renormalization factor
δm	- self-mass renormalization

The renormalization constants appear in the expressions:

$$S_F(p) = Z_2 S'_F(p)$$
$$D_F(q)_{\mu\nu} = Z_3 D'_F(q)_{\mu\nu}$$
$$V_\mu(q', q) = Z_1^{-1} V'_\mu(q', q)$$

where $S'_F(p)$, $D'_F(q)_{\mu\nu}$, and $V'_\mu(q', q)$ are the divergence-free QED physical propagators and vertex.

In addition the renormalized charge e satisfies

$$e_0 = Z_1 e/(Z_2 Z_3^{1/2}) = Z_3^{-1/2} e \qquad (4.1)$$

where e_0 is the bare electric charge and e is the renormalized charge. The self-mass δm satisfies

$$m = m_0 + \delta m$$

where m_0 is the bare electron mass and m is the physical electron mass.

We now leave the conventional discussion of QED since it is discussed in detail in many textbooks and papers.

4.3 Johnson-Baker-Willey (JBW) Massless QED Model

The JBW model was an attempt to eliminate the divergences of QED at very high energies where the electron mass may be neglected. It was extended to allow for non-zero electron mass. The model is summarized in Phys. Rev. **D8**, 1110 (1973) which is reproduced for the reader in Appendix C. This paper and references therein present a detailed derivation of the JBW model. The interested reader is referred to the JBW papers, which will only be utilized here for the discussion of the JBW eigenvalue function and its role in eliminating QED divergences as well as in calculating the fine structure constant usually denoted α.

In addition to the JBW series of papers Adler[22] made a significant advance in the understanding of the JBW model and massless QED by showing that the JBW eigenvalue function, if it has a zero at the measured value of α or at α_0 - the bare value of α, will be an essential singularity.

Subsequently this author[23] found an approximation that enabled the JBW eigenvalue function to be calculated to all orders in α. The approximate eigenvalue function did not have an essential singularity although it had poles and branches. It also did not have a zero at α near the experimentally found value of 1/137.035999139 (31) at $Q^2 = 0$. These apparent failures, and the unification of QED and the Weak interactions in the Weinberg-Salam Model, eliminated the interest in the massless QED at that time.

In this book we revive the eigenvalue condition, in a modified form, for divergence-free QED and show that it has a zero near the known value of α. We explain the absence of the essential singularity as due to our approximation. However the approximation does suggest the possibility of an essential singularity in the exact solution. On this basis we show important possible corollary effects of the eigenvalue function zero. And we then extend our discussion to the Weak and Strong interactions

---

22 S. Adler, Phys. Rev. **D5**, 3021 (1972).

23 S. Blaha, Phys. Rev. **D9**, 2246 (1974). Reprinted in Appendix A for the reader's convenience.

where we determine that eigenvalue functions exist that pin down the values of their coupling constants to values near the currently known values.

Thus we have a viable method of determining, at least approximately, the known coupling constants of the Unified SuperStandard Theory.

5. QED Eigenvalue Function

5.1 Origin of the Eigenvalue Function

The massless QED eigenvalue function of the JBW model was found in a series of papers and summarized in some detail by the paper of K. Johnson and M. Baker,[24] Phys. Rev. **D8**, 1110 (1973) in Appendix C. In this section we briefly outline the steps leading to the JBW eigenvalue function:

1. The electron self-energy and the photon vacuum polarization are calculated using the free electron and photon propagators from the Feynman diagrams of Figs. 2 and 3 in Appendix C.

2. The apparent quadratic divergence appearing in the diagrams of Fig. 3 is reduced to a logarithmic divergence in the vacuum polarization.

3. The logarithmic divergence in the vacuum polarization is further reduced to a single power of the logarithm of the ultraviolet cutoff denoted Λ. (Eq. 2.9 of Appendix C)

4. The coefficient of the divergent logarithmic term is denoted

$$(x/2\pi)F(x)$$

where x is the bare fine structure constant α_0.

[24] The author is pleased to acknowledge discussions with M. Baker and K. Johnson in 1973 that introduced him to their model.

5. If

$$F(x_0) = 0$$

   for some value x_0 then a consistent divergence-free (finite) solution of massless QED is found.

6. The F(x) function is reduced to the eigenvalue function $F_1(x)$ which is the sum of logarithmic divergences of all single closed electron loop diagrams. If the eigenvalue function $F_1(x)$ has a zero at x_0 then $F(x_0) = 0$. Consequently the eigenvalue condition becomes

$$F_1(x_0) = 0 \qquad (5.1)$$

7. Adler[25] made the important observation that a zero in eq. 5.1 would necessarily be an essential singularity:

$$d^nF_1(x)/dx^n|_{x=x_0} = 0 \qquad \text{for all } n > 0 \qquad (5.2)$$

   Depending on the summation in perturbation theory of the relevant vacuum polarization diagrams might may occur at the bare coupling constant α_0 or the physical coupling constant $\alpha = 1/137$ …

8. This author (See Appendix A.) calculated an approximation to F_1 which did not explicitly display an essential singularity and did not have a zero at the physical fine structure constant α.

5.2 Blaha's Approximate Calculation of the Eigenvalue Function

In 1974 this author[26] formulated an approximation to the equations of massless QED and solved them for the vacuum polarization, electron self-energy and the vertex

---

[25] Adler *op. cit.*
[26] Blaha *op. cit.*

renormalization. The approximation is described in detail in the author's Phys. Rev. D paper in Appendix A.

The approximate solution for $F_1(x)$ had the encouraging feature that it reproduced the known[27] low order exact calculations of $F_1(x)$:

$$F_{1\text{ low order}}(x) = 2/3 + x/(2\pi) - (1/4)[x/(2\pi)]^2 \qquad (5.3)$$

Our approximate solution, which summed pieces of the vacuum polarization given by the diagrams of Figs. 2 and 3 in Appendix A, yields the algebraic equations:[28]

$$A_1 = (g + 1)(1 - 2g^2)/[(g + 2)(g - 1)] \qquad (5.4)$$

$$A_2 = [8g^2(2g + 1) - (2g^3 + 2g^2 + g - 2)(g^2 + 2g + 2)]/[2(g^2 - 1)(g^2 - 4)]$$

$$A_3 = -2(1 + 3g + 6g^2 + 2g^3)/[g(g + 1)]$$

$$A_4 = -(g + 2)(1 + 5g + 6g^2 + 2g^3)/[g(g^2 - 1)] - 1/(g + 1)$$

$$\psi = [gA_3 - (4 + 2g)A_1]/[(4 + 2g)A_2 - g\,A_4]$$

$$(\alpha/2\pi) = [gA_4 - (4 + 2g)A_2]/(A_4A_1 - A_2A_3)$$

$$F_1(g) = (2/3)(1 - 3g^2/2 - g^3) - (\alpha/4\pi)[(2 + 4g + 4g^2)(g - 2) + \alpha\psi g^3]/[(g^2 - 1)(g - 2) + \\ + \alpha(2 + 4g + 4g^2)(g - 2) + \alpha\psi g^3]$$

as a function[29] of g with ψ specifying the gauge, and with the definitions

$$\Gamma_\mu(p) = f(\gamma_\mu + 2g\gamma\cdot pp_\mu/p^2)(p/\Lambda)^{2g} \qquad (5.5)$$

---

[27] J. Rosner, Phys. Rev. Lett. **17**, 1190 (1966).

[28] Blaha *op. cit.* The solution for the eigenvalue function is clearly best expressed in terms of the g factor in the exponents of the divergent renormalization factors.

[29] We use $F_1(g)$ and $F_1(\alpha(g))$ interchangeably.

$$S_F = [f\gamma\cdot p(p/\Lambda)^{2g}]^{-1} \tag{5.6}$$

$$\Gamma_{\mu\alpha}(p) = (f_3/p^2)(\gamma\cdot p\gamma_\mu\gamma_\alpha - \gamma_\alpha\gamma_\mu\gamma\cdot p)(p/\Lambda)^{2g} \tag{5.7}$$

and

$$F_1 = (2/3)(1 - 3g^2/2 - g^3) - f_3/f \tag{5.8}$$

in the notation of Appendix A. Eqs. 5.4 and 5.5 manifestly cannot lead to a form of F_1 with an essential singularity due to their algebraic form.

The plot of F_1 below did not show a zero of F_1 at the physical fine structure constant. Thus the hopes raised by the JBW model seemed dashed—at least in our approximate solution *then*. Later in this book we revive the hope of a satisfactory eigenvalue function with an eigenvalue at the physical value of the fine structure constant α.

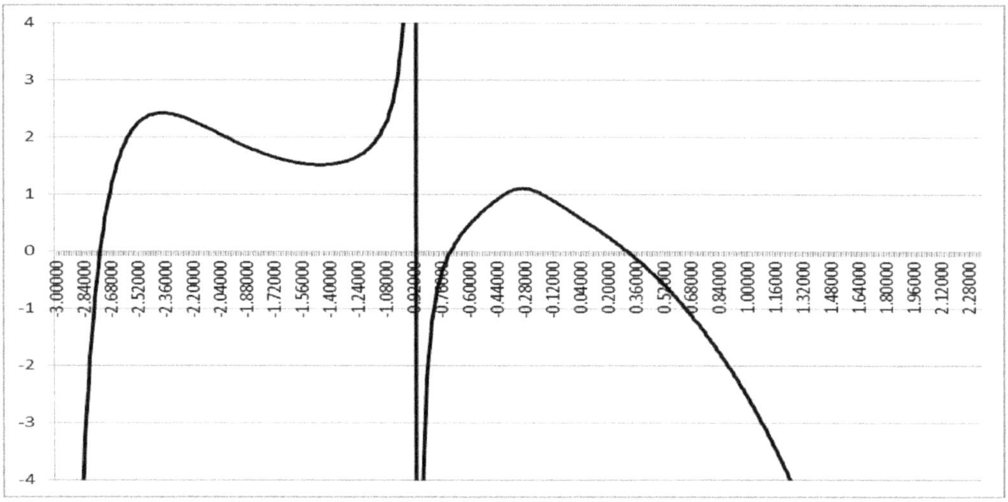

Figure 5.1. A plot of the approximate eigenvalue function $F_1(g)$ (vertical axis) *as a function of g.* Note none of its zeroes correspond to the known physical value of α. (Confirmed by next figure.) It does not have an essential singularity.

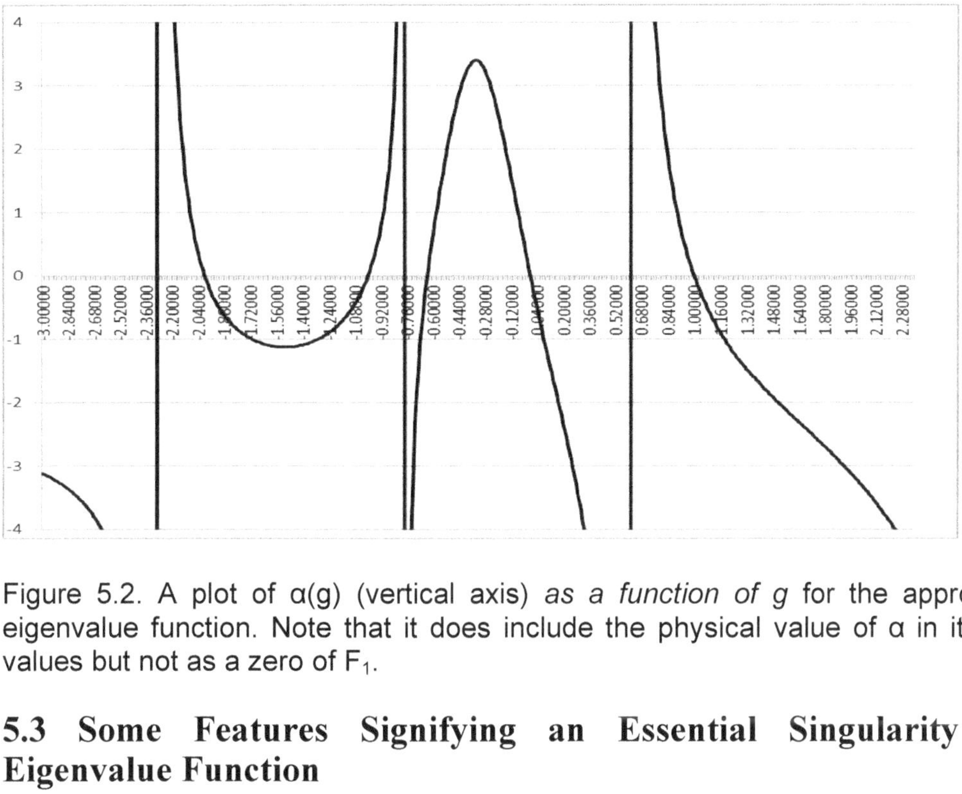

Figure 5.2. A plot of α(g) (vertical axis) *as a function of g* for the approximate F_1 eigenvalue function. Note that it does include the physical value of α in its range of values but not as a zero of F_1.

5.3 Some Features Signifying an Essential Singularity of the Eigenvalue Function

A number of important features follow primarily from the existence of an essential singularity in the JBW eigenvalue function:

1. The essential singularity may occur at the value of the bare coupling constant or at the renormalized (physical) coupling constant. This disparity is due to the appearance of differing results depending on the order of summation of perturbation theory for the vacuum polarization.

2. Due to the essential singularity, the large n coefficients ($n \geq k$) of an expansion in α of the eigenvalue function

$$F_1(\alpha) = \sum_{n=k}^{\infty} c_n \alpha^n \tag{5.9}$$

for some value k may reveal features of F_1 near the point of the essential singularity.

5.3.1 Additional Features of the Eigenvalue Function

The JBW massless QED model has one species of charged fermion. There are more charged fermion species. However their effect is simply to create a multiple of the eigenvalue function:

k Charged Fermion Species → Eigenvalue function = $kF_1(\alpha)$

where k is the number of species. Thus the eigenvalue condition is independent of number of fermion species.

The Strong interactions and other non-abelian interactions are also not necessarily relevant for the calculation of the QED F_1 and its eigenvalue since these interactions are asymptotically free and may be neglected at very high energies.

5.3.2 Recognizing Evidence In an Approximation to an Essential Singularity

In calculating a function approximately, that is known to have an essential singularity, there are several possible types of evidence of the singularity:

1. The singularity will manifest itself directly as an infinite order zero or as in infinite essential singularity value at a point.

2. Since the singularity may reveal itself only approximately, the forms of a revealing approximation may be a flat region of a function (a constant value in a region.) Or it may be an "infinite-valued" point: a pole or a singularity with a branch point.

A simple example illustrating this second possibility is the function:

$$\exp(-1/x) = 1 - 1/x + \ldots \qquad (5.10)$$

If the approximation finds only the first term of the expansion, then, since constants are trivially infinite order zeroes, a constant region in the approximation might signal an essential singularity. If the approximation has something like the first two terms in the eq. 5.10 expansion, then an infinity in the approximation could signal the presence of the essential singularity.

We shall use this working approach to detecting an essential singularity in the eigenvalue functions we study below.

5.4 New Revised Eigenvalue Function Approximation

As pointed out in our 1974 Phys. Rev. D paper The eigenvalue function F_1 does not have a zero at the known value of the Fine Structure Constant $\alpha = 1/137.035999139$ (31) at $Q^2 = 0$. It is not even close to the value of α. As a result in 1974 we abandoned the effort thinking the approximate calculation was either insufficient to capture the essential singularity or that α was possibly determined by some other, perhaps cosmological, consideration.

We now reconsider our approximation and show that item 2 of section 5.3 (eq. 5.9) leads to a value of α near to the measured value. We define

$$F_2(\alpha) = F_1(\alpha) - [2/3 + \alpha/(2\pi) - (1/4)[\alpha/(2\pi)]^2] \qquad (5.11)$$

where we subtract the known low order terms of F_1 (eq. 5.3) in accordance with eq. 5.9. $F_2(\alpha)$ will be seen below to have a neighborhood where $F_2(\alpha) \approx 0$ with a set of values for α including an approximate value for the known fine structure constant. Since our $F_1(\alpha)$ (and consequently $F_2(\alpha)$) is an approximate solution of the single electron loop QED equations, it is reasonable to expect $F_2(\alpha)$ is not identically zero at the eigenvalue point. However, a possible indication of an essential singularity is a *constant region* of $F_2(\alpha)$, as noted in section 5.3.2 using the example eq. 5.10. *We find such a region exists of constant $F_2(\alpha)$. That region has a value at its "midpoint" closely approximating the known value of α.*

In terms of F_2 the renormalization constant Z_3 is

$$Z_3 = 1 + F_1(\alpha)\ln(p/\Lambda) = 1 + F_2(\alpha) + \text{divergent terms} = 1 + \text{divergent terms} \quad (5.12)$$

5.5 Revised JBW Model Goal

The original goal of the JBW Model was to solve massless QED in a manner that made all renormalization constants either 1 or at least finite. We shall see that we can obtain an eigenvalue function F_2 that has a zero at the known fine structure constant that we denote α. Until now we have not specified the value α that appears in the preceding equations. We now define α as a partially renormalized quantity that is related to the bare fine structure constant α_0 by

$$\alpha = \alpha_0[2/3 + \alpha_0/(2\pi) - (1/4)[\alpha_0/(2\pi)]^2] \quad (5.13)$$

and specify all appearances of α in eqs. 5.4-5.11 as the α in eq. 5.13 which we will show leads to the approximate physical value of the fine structure constant: $\alpha = 0.007297354$.

Thus the renormalized expressions appearing in eqs. 5.5 - 5.7 are not fully finite. However the intermediate renormalized finite α is physically sensible—more so than the completely finite renormalization constants goal of the JBW Model. The bare charge constant α_0 does approach ∞ at very short distances. The simplest examples of this phenomenon are the physical Coulomb scattering amplitudes and the first order change in hydrogen-like atomic energy levels.[30] Thus our modified JBW Model with a partial renormalization (eq. 5.13) conforms to reality.

$$Z_3 = 1 + \{\alpha F_2(\alpha) + \alpha[2/3 + \alpha/(2\pi) - (1/4)[\alpha/(2\pi)]^2]\}\ln(p/\Lambda) \quad (5.14)$$
$$= 1 + \alpha\{2/3 + \alpha/(2\pi) - (1/4)[\alpha/(2\pi)]^2\}\ln(p/\Lambda)$$

at α = the physical fine structure constant where $F_2(\alpha) = 0$.

5.5.1 Physical Implications of Intermediate Renormalization

There is no problem with using F_2 as the eigenvalue function since it can be made the factor relating an infinite bare charge to the physical charge (eq. 4.1). (It is an

[3030] See E. A. Ueling, Phys. Rev. **48**, 55 (1935) and R. Serber, Phys. Rev. **48**, 49 (1935).

imperfection viewed from the goal of the original JBW model for a divergence-free QED.) However we believe our modified model is reality.

The below plots show the overall form of $F_2(\alpha)$ plotted vs. g and the region where the fine structure value appears. Under these circumstances it appears that the Modified JBW model of QED is correct and determines the fine structure constant if it could be completely solved. *As a result other proposed determinations for the value of α appear to be ruled out: including cosmological and other physical approaches.* Thus the fundamental nature of the universe, and of Life, is fixed by QED.

5.5.2 Intermediate Renormalization of Weak and Strong Coupling Consttants

As we noted above in our discussion of the first order change in hydrogen-like atomic energy levels, the bare electron charge e_0 is changed by $\sqrt{Z_3}$ vacuum polarization. The proton bare charge, also e_0, is similarlychanged by $\sqrt{Z_3}$ so the charge factor in the level shift is the renormalized charge $e_0^2Z_3$. Since protons are composed of three quarks, their charges must also be changed by a factor of $\sqrt{Z_3}$. Thus they have infinite bare electric charges.

5.5.3 Summary of QED Eigenvalue Results

In Fig. 5.4 below we find that at the mid-point of the "constant" region where

$$g = -0.000580537$$

that

$$\alpha_{calculated}(g) = 0.007297354$$

with

$$\alpha_{calculated}(g)^{-1} = 137.0359801$$

compared to the actual fine strcture constant value

$$\alpha = 0.007297353$$

with

$$\alpha^{-1} = 137.0359991$$

displaying a very close match.[31]

The negative value of the exponential factor g above indicates that Z_3 diverges as $\Lambda \rightarrow \infty$ showing QED is not asymptotically free and is divergent in our approximation. (See eqs. 5.5 – 5.7.)

5.6 Plots of the New Eigenvalue Function Approximation

In this section we display plots of $F_2(\alpha)$ that show $F_2(\alpha) \approx 0$ in a region of α values, in the center of which the value of the fine structure constant appears.

[31] A very small change in the value of g would give an exact match.

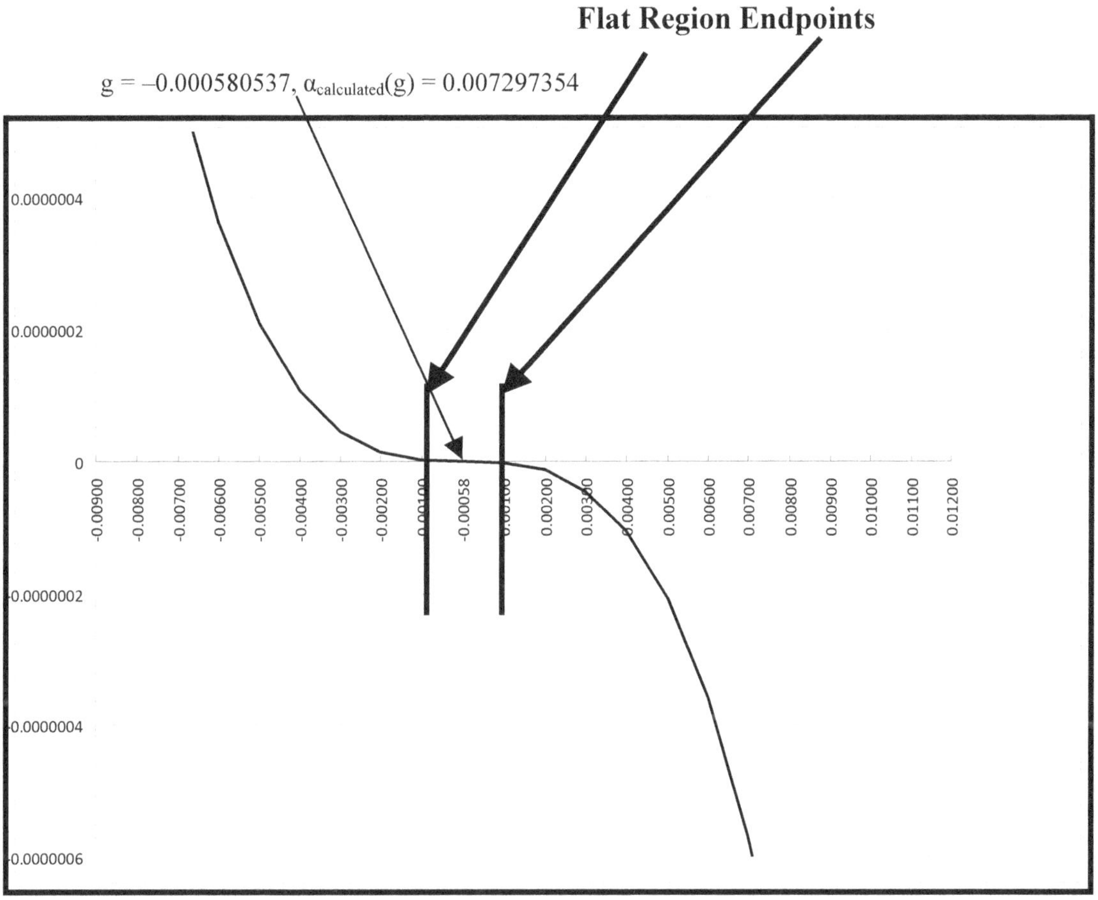

Figure 5.3. Plot of revised eigenvalue function $F_2(g)$ (vertical axis) vs. g. Note the central point of the "flat region" corresponds very closely to the approximate value of α.

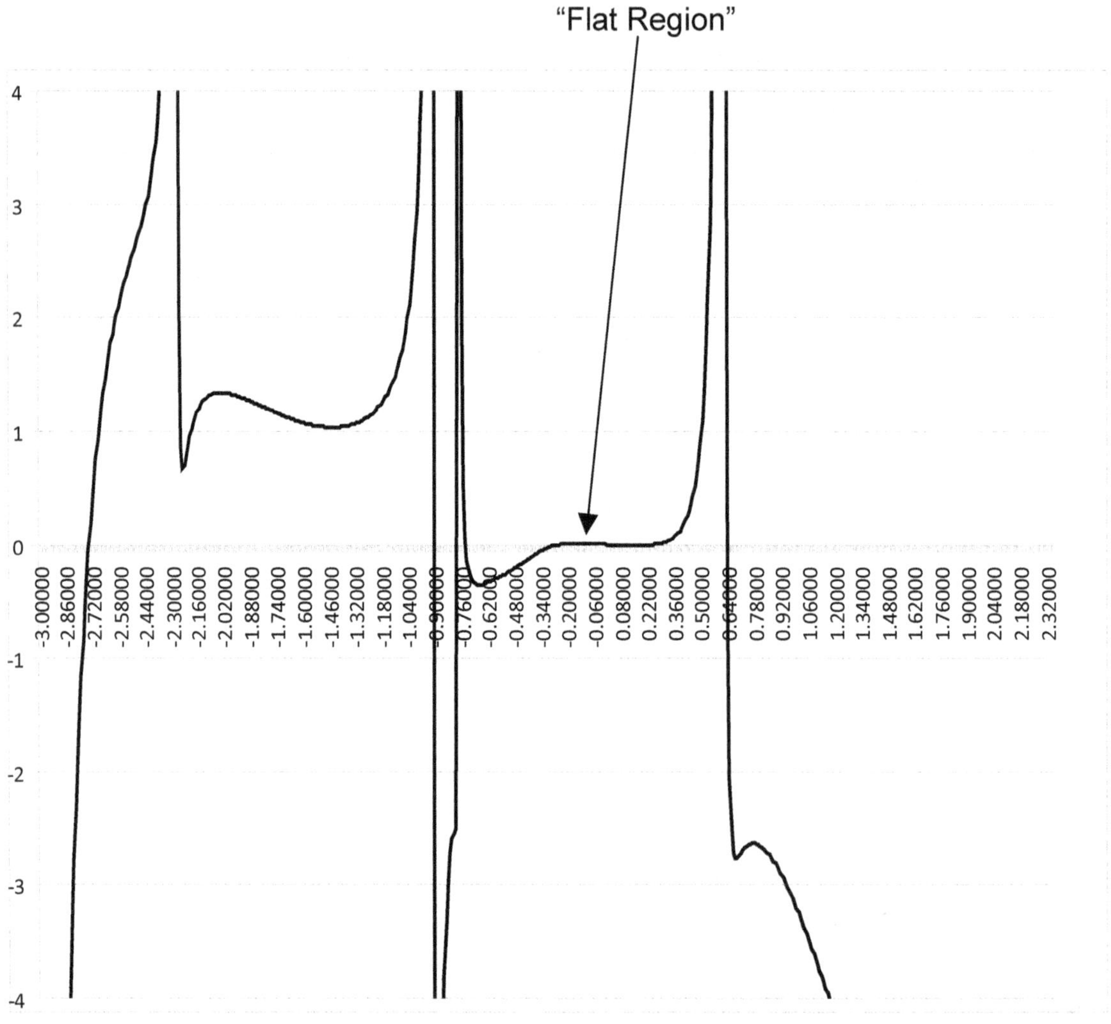

Figure 5.4 Plot of the fine structure "eigenvalue" region of $F_2(g)$ (vertical axis) with α values displayed. The midpoint of the "flat" region is the point with the values of g and α shown above.

6. Non-Abelian Interactions "Fine Structure Constants"

The Unified SuperStandard Theory has non-abelian interactions that go beyond QED by having group transformations that necessitate the introduction of cubic and quartic terms in the lagrangian of the theory. The calculation of non-abelian coupling constants have until now been limited to low order perturbative running coupling constant calculations.[32]

It would have been better to have calculations of coupling constants to all orders—even if only approximately—as we successfully saw in massless QED earlier. In this chapter and the following chapter we will attempt to calculate coupling constants in a manner analogous to that of earlier chapters for QED. Our attempt will be at very high energies where we can assume that all particle masses are negligible. We will make a number of assumptions which are not unreasonable:

1. We will assume that infrared divergences are not relevant.

2. We will assume that cubic and quartic Yang-Mills couplings can be neglected reducing Yang-Mills fields to multiplets of QED-like fields.

3. We will assume that each non-abelian interaction can be considered independent of all other interactions.

4. We will assume that each interacting field of a group is subject to the same vacuum polarization renormalization. This renormalization is assumed to be similar to that of JBW vacuum polarization.

---

[32] See, for example, H. Georgi, H. R. Quinn, S. Weinberg, Phys. Rev. Lett. **33**, 451 (1974).

5. Thus we assume that an eigenvalue function exists for each non-abelian interaction that results from a single logarithmic "divergence" in the vacuum polarization. This divergence stems from one fermion loop vacuum polarization diagrams summed over all fermion species of the fundamental representation of its group.

The justification for this series of assumptions is the similarity of the contributions of the vacuum polarization diagrams if cubic and quartic non-abelian interactions are neglected. The calculations of each Feynman diagram have the same characteristics of massless QED momentum space integrals up to an overall interaction constant factor in each order of perturbation theory.

Based on these assumptions we proceed to calculate non-abelian eigenvalue functions for SU(2), SU(3), and SU(4) in the following chapter.

We do make one further assumption we assume that the coupling constant for each interaction has the factor

$$c_G^{-1} = [(11/3)C_{ad} - 2C_f/3]/(16\pi)^3 \qquad (6.1)$$

to all orders where C_{ad} is the dimension of the fundamental representation of the non-abelian group and C_f is the number of fermions (fermion flavor) of the interaction.

The eigenvalue function F_1 for an interaction with coupling constant α can be expressed as a power series in α:

$$F_1(\alpha_G) = \sum_n a_n(\alpha_g)^n = \sum_n a_n(c_G\alpha)^n \qquad (6.2)$$

under the assumption that the interaction group constant is approximately c_G to all orders in n where the QED eigenvalue function has the form

$$F_1(\alpha) = \sum_n a_n\alpha^n \qquad (6.3)$$

Since non-abelian interactions are known to be asymptotically free—with coupling constants becoming finite at ultra-short distances we will not need to do intermediate renormalization such as we did for electric charge. Thus we will use F_1 as the eigenfunction.

Having established the framework for our approximate calculations of the eigenvalue functions and their eigenvalues we will proceed to calculate them in the next chapter for SU(2), SU(3) and SU(4).

Note that the procedure for obtaining the massless QED eigenvalue function and the fine structure constant will exactly duplicate our calculation in Appendix A and chapter 5.

Chapter 11 provides a simple fit to the F_2 and α functions using tangents.

7. Non-Abelian "Fine Structure Constant" Eigenvalue Conditions

7.1 Coupling Constants for Non-Abelian Interactions

The vector interactions and coupling constants of the Unified SuperStandard Theory are:[33]

- The strong interaction coupling constant[34] $g_S = 1.22$
- The Weak SU(2) coupling constant $g_W = 0.619$
- The Electromagnetic U(1) coupling constant $e = g_E = 0.303$
- The Dark Weak SU(2) coupling constant $g_{DW} = ?$
- The Dark Electromagnetic U(1) coupling constant $g_{DE} = ?$
- The U(4) Generation group coupling constant $g_G = ?$
- The U(4) Layer group coupling constant $g_L = ?$
- The U(4) Species group coupling constant $g_{Sp} = ?$
- The U(192) Θ-interaction group coupling constant $g_\Theta = ?$

---

[33] All coupling constant values are based on data extracted from C. Patrignani *et al* (Particle Data Group), Chinese Physics **C40**, 100001 (2014).

[34] Based on the running coupling constant value $\alpha_s(M_Z^2) = 0.1193 \pm 0.0016$.

In this chapter we will calculate (approximately) g_S, g_W, e, and the "fine structure constant" for SU(4) g_4, which would seem to apply to the U(4) groups listed above, subject to the approximations listed in chapter 6.

We will use the below to calculate "fine structure constant"s α_G

$$A_1 = (g + 1)(1 - 2g^2)/[(g + 2)(g - 1)] \tag{7.1}$$

$$A_2 = [8g^2(2g + 1) - (2g^3 + 2g^2 + g - 2)(g^2 + 2g + 2)]/[2(g^2 - 1)(g^2 - 4)]$$

$$A_3 = -2(1 + 3g + 6g^2 + 2g^3)/[g(g + 1)]$$

$$A_4 = -(g + 2)(1 + 5g + 6g^2 + 2g^3)/[g(g^2 - 1)] - 1/(g + 1)$$

$$\psi = [gA_3 - (4 + 2g)A_1]/[(4 + 2g)A_2 - g\, A_4]$$

$$(\alpha_G/2\pi) = c_G^{-1}[gA_4 - (4 + 2g)A_2]/(A_4A_1 - A_2A_3) \tag{7.2}$$

$$F_1 = (2/3)(1 - 3g^2/2 - g^3) - (\alpha_G/4\pi)[(2 + 4g + 4g^2)(g - 2) + \alpha_G\psi g^3]/[(g^2 - 1)(g - 2) + $$
$$+ \alpha_G(2 + 4g + 4g^2)(g - 2) + \alpha_G\psi g^3] \tag{7.3}$$

using eq. 6.1.

7.2 SU(2) ElectroWeak "Fine Structure Constant" Eigenvalue Function

The U(1) QED eigenvalue function and eigenvalue was discussed in chapter 5. In this section we discuss and show plots of the SU(2) ElectroWeak sector eigenvalue function subject to the discussion in chapter 6 and the above. In particular we calculate the SU(2) eigenvalue function and eigenvalues using eqs. 7.2 and 6.1.

Due to the asymptotic freedom of non-abelian interactions we anticipate that the value of the exponential factor g (eqs. 5.5 – 5.7) will be positive at the "fine structure constant" eigenvalue of the SU(2) eigenvalue function which we take to have the form shown above due to the assumptions of chapter 6.

We define $F_{1su(2)}(\alpha)$ using eq. 7.3 above with $c_{GSU(2)}^{-1}$ = -0.003023589 from eq. 6.1 using $C_{ad} = 2$ and $C_f = 2$.

In our discussion of QED in chapter 5 we found a negative value for g and a consequent divergence in Z_3 as $\Lambda \to \infty$. We then found that $F_1(\alpha)$ did not yield the QED eigenvalue. We had to introduce $F_2(\alpha)$ by eliminating low order (in α) terms using intermediate renormalization. Those terms have logarithmic divergences. Their elimination did not create a problem since the QED renormalizations also diverge due to $g < 0$ at the eigenvalue point. The total divergence is generated by combining the divergent factors due to the $(p/\Lambda)^{2g}$ factor in Z_3 with the omitted divergent terms of F_1. See eq. 5.12.

In the case of non-abelian interactions $g > 0$, which implies the renormalizations go to 1 as $\Lambda \to \infty$ (asymptotic freedom).Thus we will not truncate F_1 but will calculate using it (eq. 7.3) to find an approximation to the "fine structure constant" eigenvalues of non-abelian interactions. *As $\Lambda \to \infty$ non-abelian theories approach free field theories.*

The signature of the eigenvalue function essential singularity in the present case will be a divergence in F_1 for $g > 0$. It signals that we have an approximation to the expected essential singularity. See section 5.3.2.

The below plots show the overall form of $F_{1SU(2)}(\alpha)$ plotted vs. g and the divergent region of $F_{1SU(2)}(g)$ specifying the "fine structure constant" eigenvalue. Our approximate "fine structure constant" values in this case and the SU(3) and SU(4) cases are quite reasonable. We find the "essential singularity" point at

$$g = 0.54$$

where $\alpha_{SU(2)} = g_W^2/(4\pi)$

$$\alpha_{calculatedSU(2)}(g) = 0.0425$$

compared to the actual measured "fine structure constant" value

$$\alpha_{SU(2)} = 0.0305$$

displaying a fairly close match given the approximate nature of the calculations.The positive value of the exponential factor g above indicates that $Z_3 \to 1$ as $\Lambda \to \infty$ showing the SU(2) interaction is asymptotically free. (See eqs. 5.5 – 5.7.) It is interesting to note that the “fine structure constant” plot has a divergence at g = 0.58.

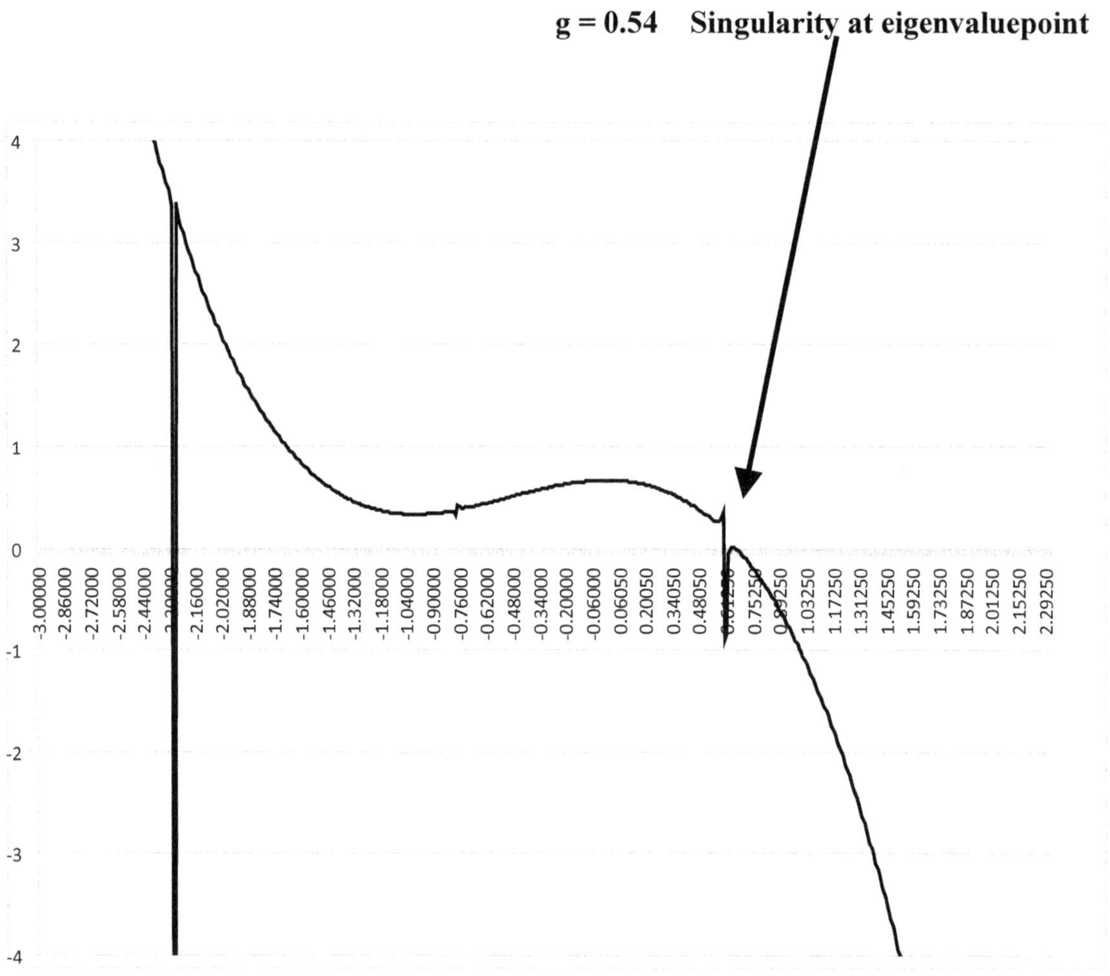

Figure 7.1. Plot of $F_{1su(2)}$ (vertical axis) as a function of g.

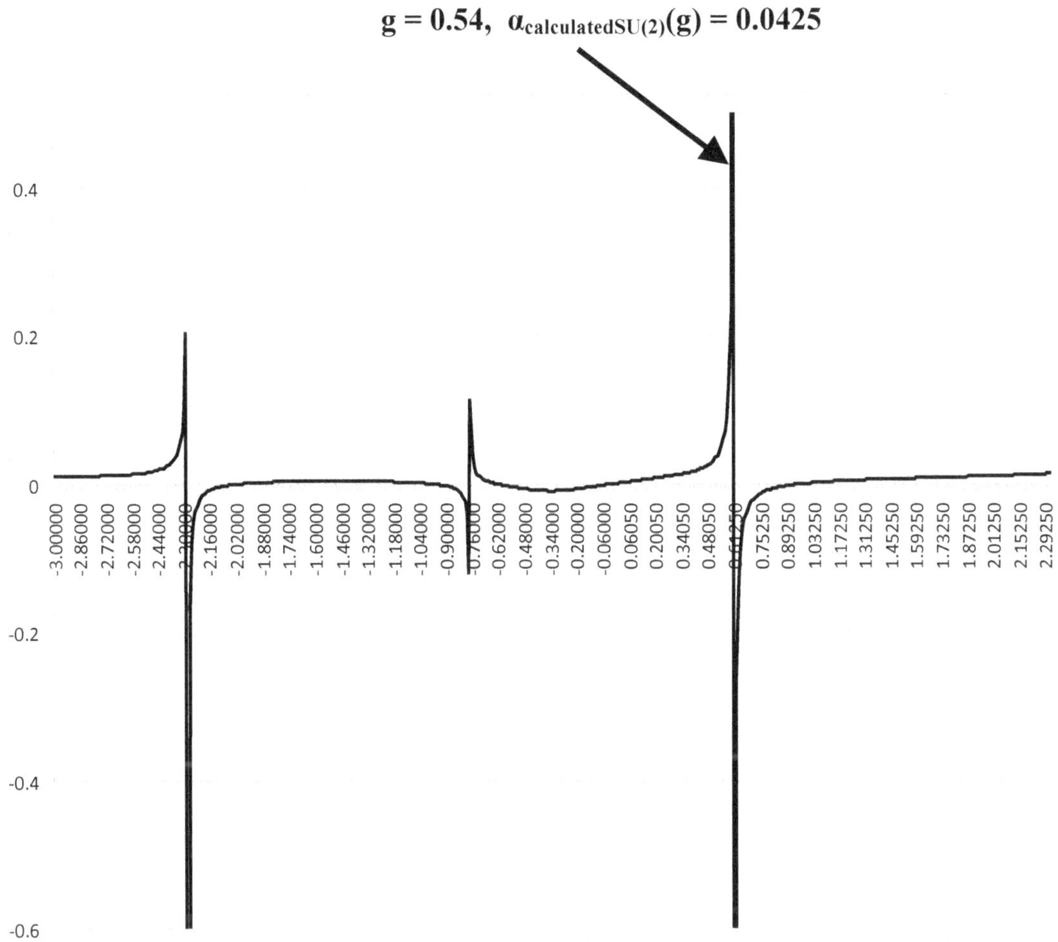

Figure 7.2. Plot of the "fine structure constant" $\alpha_{calculatedSU(2)}$ (vertical axis) as a function of g.

7.3 SU(3) Strong "Fine Structure Constant" Eigenvalue Function

The discussion of the SU(3) Strong interaction case is very much the same as the preceding SU(2) discussion. We define $F_{1su(3)}(\alpha)$ using eq. 7.3 above with

$$c_{GSU(3)}{}^{-1} = -0.004535384$$

from eq. 6.1 using $C_{ad} = 3$ and $C_f = 3$.

The signature of the eigenvalue function essential singularity in the present case will be a divergence in F_1 for $g > 0$. It signals that we have an approximation to the expected essential singularity. See section 5.3.2.

The below plots show the overall form of $F_{1SU(3)}(\alpha)$ plotted vs. g and the divergent region of $F_{1SU(3)}(g)$ specifying the "fine structure constant" eigenvalue. We find the "essential singularity" point at

$$g = 0.5605$$

where

$$\alpha_{calculatedSU(3)}(g) = 0.086$$

compared to the actual "measured" "fine structure constant" value (at $Q^2 = 2$ GeV)

$$\alpha_{SU(3)} = 0.118$$

again displaying a fairly close match given the approximate nature of the calculations.

The positive value of the exponential factor g above indicates that $Z_3 \rightarrow 1$ as $\Lambda \rightarrow \infty$ showing the SU(3) interaction is asymptotically free. (See eqs. 5.5 – 5.7.) It is interesting to note that the "fine structure constant" plot also has a divergence at g = 0.6105.

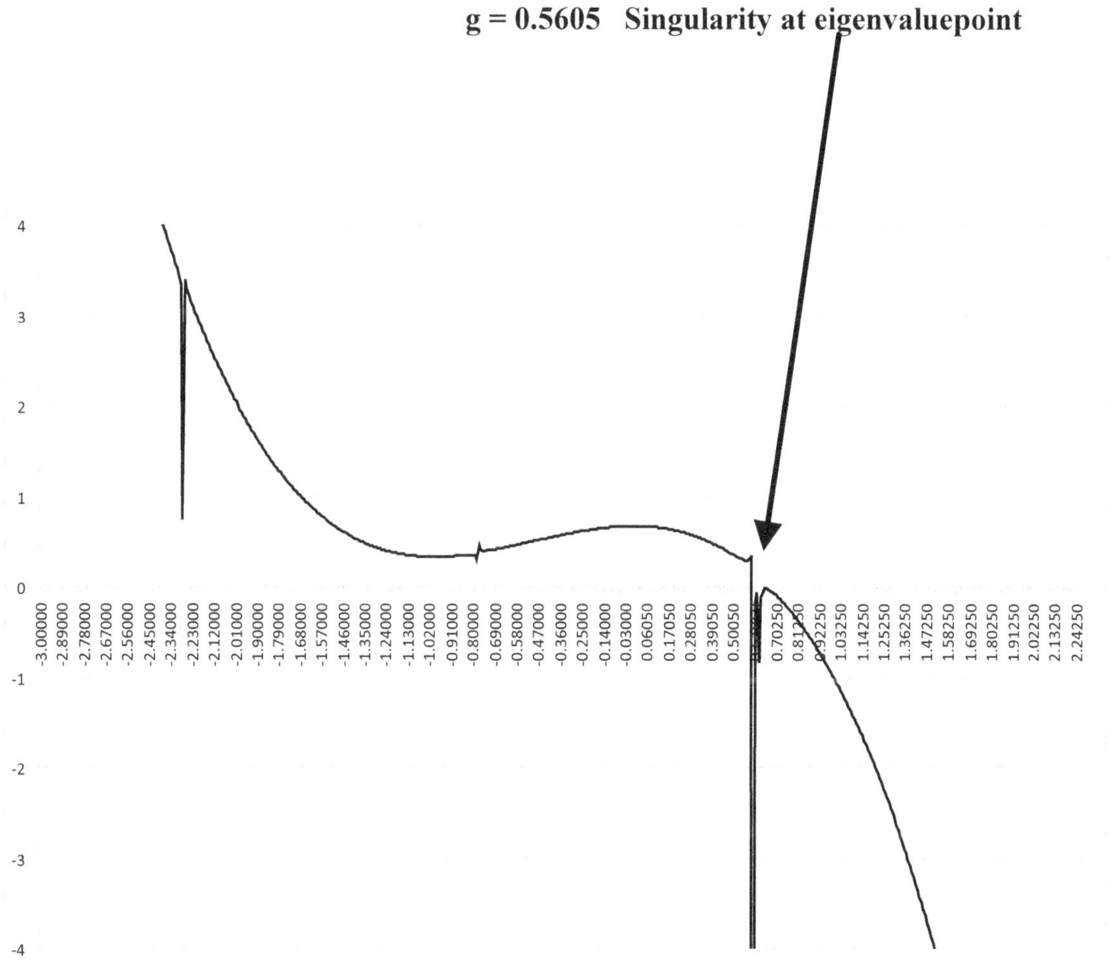

Figure 7.3. Plot of $F_{1su(3)}$ (vertical axis) as a function of g.

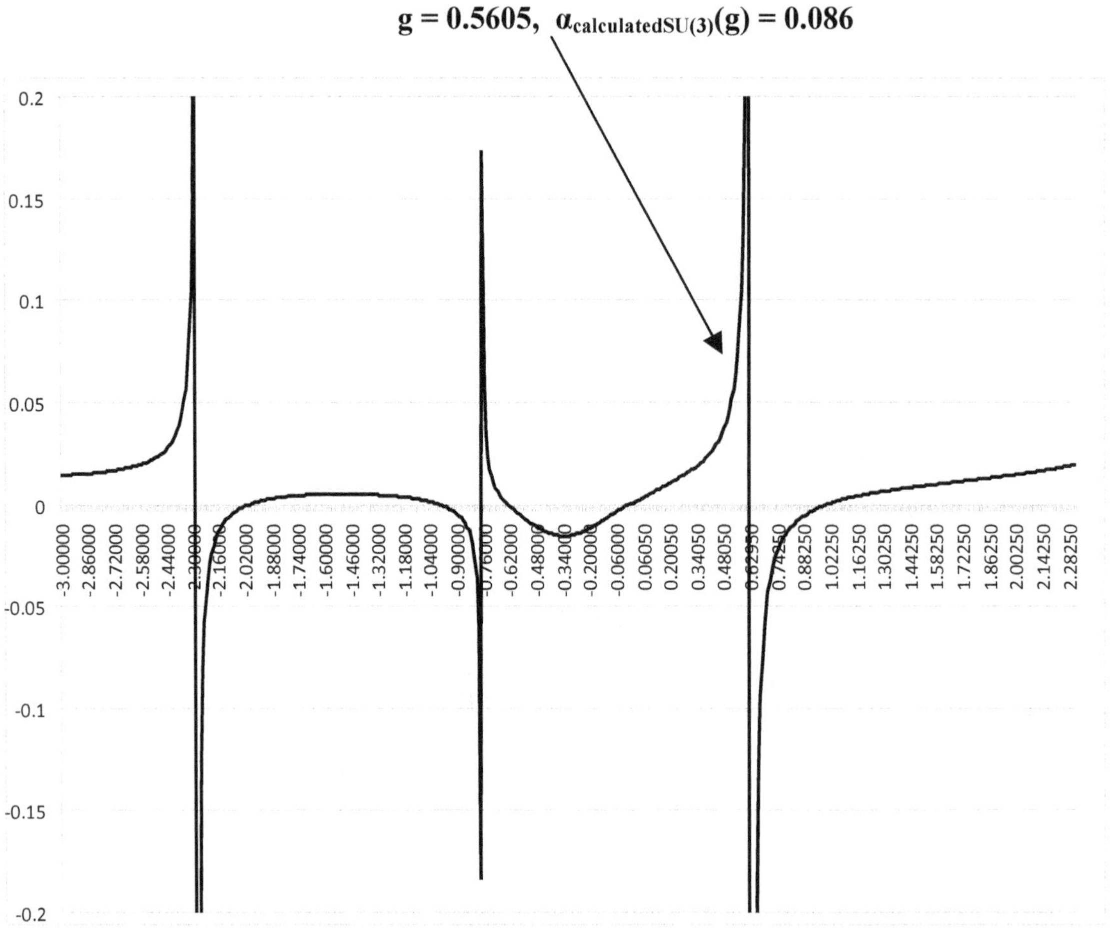

Figure 7.4. Plot of $\alpha_{calculatedSU(3)}$ (vertical axis) as a function of g.

7.4 SU(4) "Fine Structure Constant" Eigenvalue Function

The discussion of the SU(4) case of *other* interactions is very much the same as preceding discussions. We define $F_{1su(4)}(\alpha)$ using eq. 7.3 above with

$$c_{GSU(4)}^{-1} = -0.006047178$$

from eq. 6.1 using $C_{ad} = 4$ and $C_f = 4$.

The signature of the eigenvalue function essential singularity in the present case will be a divergence in F_1 for $g > 0$. It signals that we have an approximation to the expected essential singularity. See section 5.3.2.

The below plots show the overall form of $F_{1SU(4)}(\alpha)$ plotted vs. g, and the divergent region of $F_{1SU(4)}(g)$ specifying the "fine structure constant" eigenvalue. We find the "essential singularity" point at

$$g = 0.598$$

where

$$\alpha_{calculatedSU(4)}(g) = 0.384$$

compared to the conjectured[35] "fine structure constant" value

$$\alpha_{SU(4)} = 0.458$$

again displaying a fairly close match given the approximate nature of the calculations.

The positive value of the exponential factor g above indicates that $Z_3 \rightarrow 1$ as $\Lambda \rightarrow \infty$ showing the SU(4) interaction is asymptotically free. (See eqs. 5.5 – 5.7.) It is interesting to note that the "fine structure constant" plot has a divergence at $g = 0.618$.

---

[35] This value is based on the "doubling trend" seen in the three known coupling constants in chapter 8.

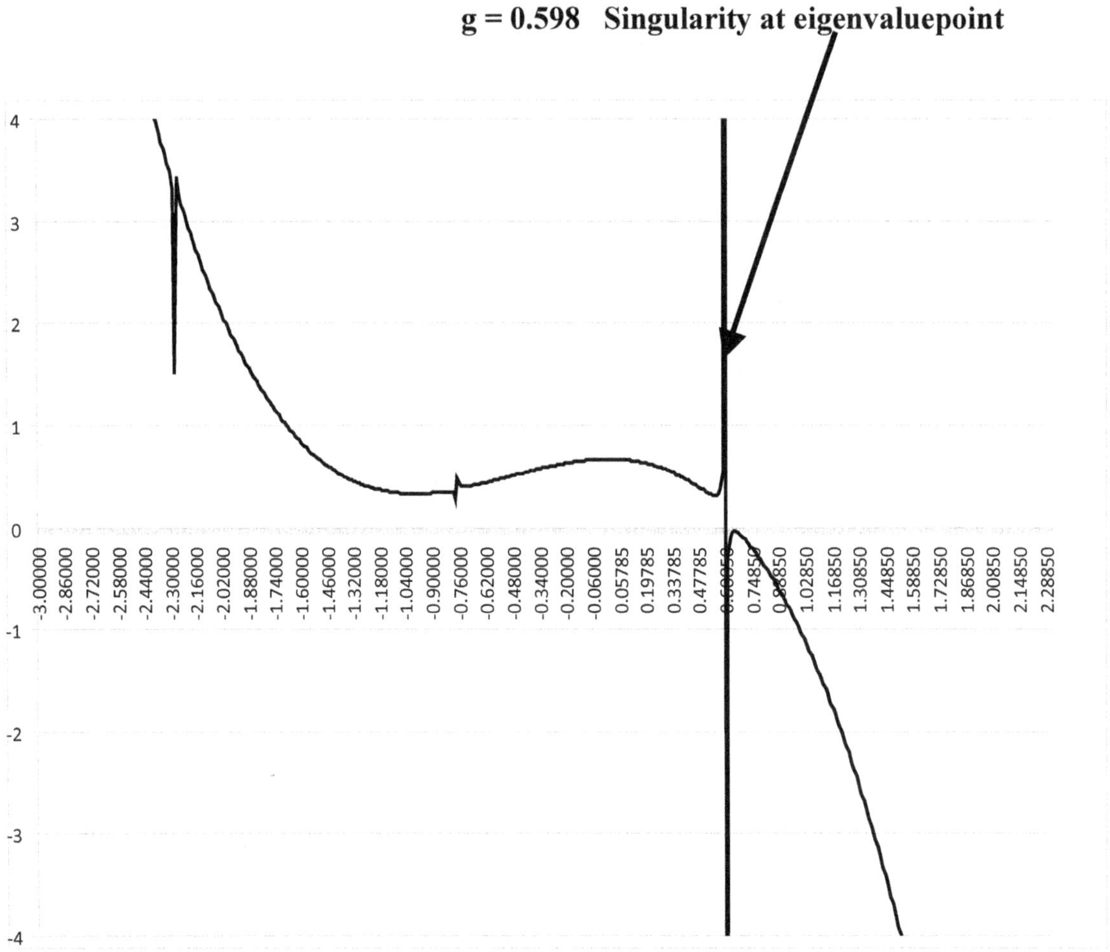

Figure 7.5. Plot of $F_{1su(4)}$ (vertical axis) as a function of g.

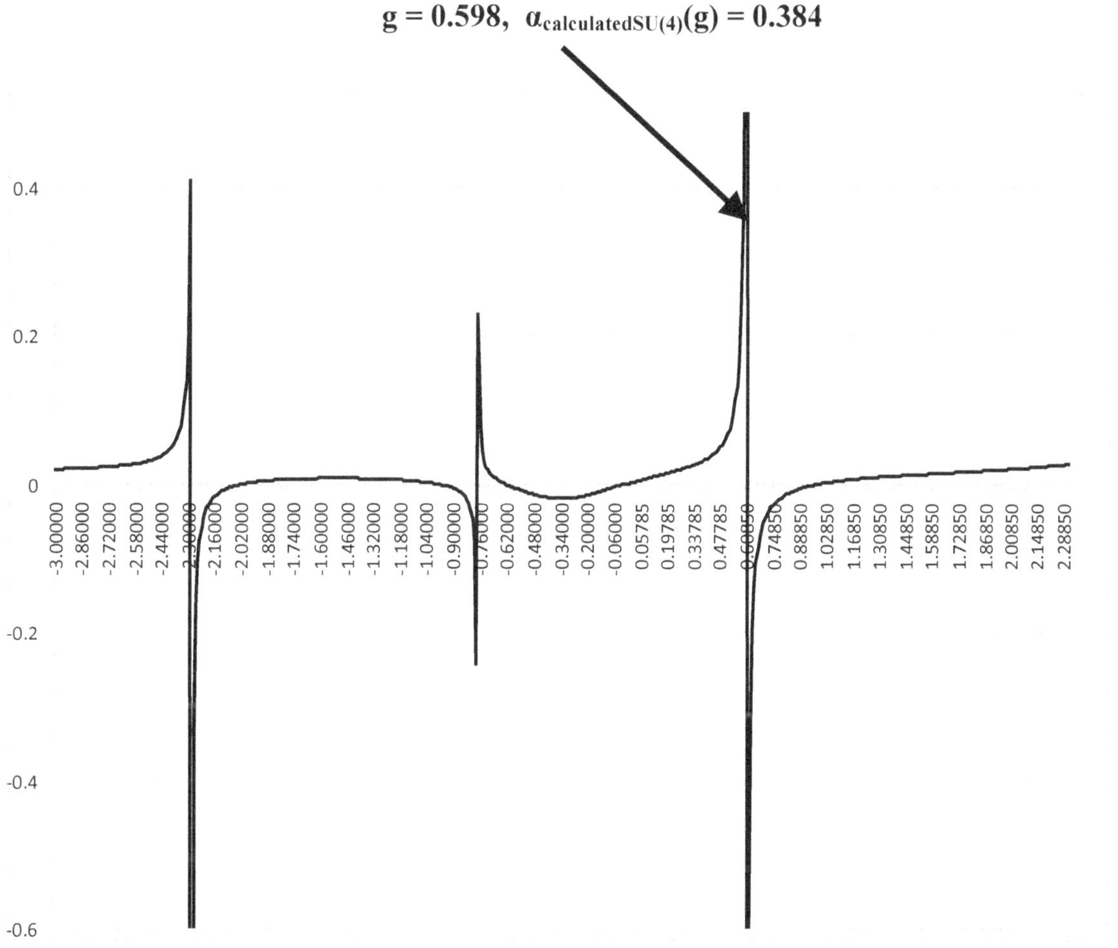

Figure 7.6. Plot of $\alpha_{calculatedSU(4)}$ (vertical axis) as a function of g.

8. "Fine Structure Constants" of the Unified SuperStandard Theory

In this chapter we summarize the values of the approximate values of the α_G determined in the previous chapter.

Group	*Known Coupling Constant*	*Known* $g_G^2/(4\pi)$	*Calculated* $\alpha_G = g_G^2/(4\pi)$	*Calculated*[36] *Exponent* g_G
QED, U(1)	0.303	$\alpha^{-1} = 137.0359991$	$\alpha^{-1} = 137.0359801$	–0.000580537
SU(2)	0.619	0.0305	0.0425	0.54
SU(3)	1.22	0.118	0.086	0.5605
SU(4)	2.4?[37]	0.458?	0.384	0.598

The relative closeness of the calculated values of "fine structure constants" to the experimentally known values is very encouraging—particularly in the case of the Electromagnetic fine structure constant α. It puts to rest other possible explanations for its value.

Our QED calculation of α has no free (adjustable) parameters unlike other attempts in the past. It also is soundly based on Quantum Field Theory. The calculation of the non-abelian coupling constants also has no free (adjustable) parameters.

Thus the coupling constant eigenfunctions depend only on inherent perturbation theory based on dynamics. Coupling constant values cannot be "tweaked" to their known values by adjusting input parameters.

The ability of our 1973-4 calculation of the JBW eigenvalue function together with the new insights into understanding of the precise method to obtain its "fine

[36] They appear in eqs. 5.5 – 5.8 above.

[37] This value is based on the "doubling trend" seen in the three known coupling constants above.

structure constant" eigenvalues is also encouraging. It opens the possibility that the Unified SuperStandard Theory has within itself the mechanism for determining the constants appearing within it. It raises the hope that a similar self-determination mechanism may also exist within the theory to determine the masses appearing in the Higgs particles sector of the theory.

The result would be a self-contained all-encompassing fundamental theory—the Holy Grail of fundamental Physics.

9. Possible Exact Form of the Non-Abelian Eigenvalue Eigenfunctions

Figs. 7.4-7.6 of chapter 7 display a repetitive pattern that is similar to those of the trigonometric functions. In particular they show a similarity to the tangent function. Our approximate F_1 eigenvalue functions for non-abelian interactions, plotted in these figures, is, in each case, an approximation to the sum of one loop vacuum polarization diagrams that we calculated in our paper in Appendix A in 1974.

In this chapter we further approximate F_1 for the purpose of studying non-abelian interaction running coupling constants in chapter 10.

Since the F_1 eigenvalue functions that we calculated in chapter 7 yielded good approximate values for the Weak and Strong coupling constants, it seems reasonable to believe that these eigenvalue function approximations are close to the real F_1 eigenvalue functions that would have been obtained in precise calculations summing one fermion loop Feynman diagrams.

It is also reasonable to believe that the true F_1 functions are less complicated than the approximate ones. Some past perturbation theory summations to all orders have shown a remarkable simplicity in the resulting summation. A particular example is the author's leading logarithm summation for the deep inelastic e-p structure functions.[38] In this paper a remarkable cancellation of diagrams, due to a Stirling Numbers of the Third Kind identity,[39] led to a simple compact result.

In the present case we will take the periodic pattern in the approximate F_1 functions to be tangent functions indications in the exact F_1 eigenfunctions. We view our approximate F_1 eigenfunctions as generated by a subset of the total one loop

[38] Stephen Blaha, Phys. Rev. D **3**, 510 (1971). This paper (the author's Ph.D. Thesis) showed that perturbation theory could not account for deep inelastic e-p scaling—an open question at the time.

[39] This type of Stirling number had not been encountered in perturbation theory before, or after, the author's paper.

contributions to the "vacuum polarization" of the non-abelian interaction coupling constants.

9.1 "Exact" Form of Non-Abelian Eigenvale Functions

We suggest the correct form of the F_1 vacuum polarization eigenfunctions for the group G is

$$F_{G1}(g) = \tan[\pi(g + d_{Gf})/d_{Gd}] \qquad (9.1)$$

and the coupling constant function is the absolute value[40]

$$\alpha_G(g) = |c_G \tan[\pi(g + d_{G\alpha})/d_{Gd}]| \qquad (9.2)$$

For some value of g, $F_{G1}(g_0) \rightarrow \infty$ and $\alpha_G(g_0)$ is the coupling constant for interaction group G. We approximate the quantity d_{Gd} by $d_{Gd} = 1.29911 - 0.08929g$. Higher powers of g would be required to get a better fit—as we shall see in the following figures. We leave that issue to future work. The quantities d_{Gf} and $d_{G\alpha}$ are assumed to be constants. The graphs below show a good approximation of tangent fits to the approximate F_1 plots.

9.2 Similarity to the Madhava-Leibniz Representation of π

The form of the coupling constant eigenvalue function is comparatively simple compared to exact expressions found earlier. Remarkably it also suggests that α has a representation similar in character to the Madhava-Leibniz representation of $\pi = 3.14159...$:

$$\pi = 4 \arctan(x) \qquad (9.3)$$

for x = 1. Note the eigenvalue implied by eq. 9.2 is

[40] The absolute value is physically required to maintain the reality of coupling constants. Eq. 9.2 is an approximation that does not exclude imaginary coupling constants.

$$\alpha_G = |c_G \tan(x)| \tag{9.4}$$

(in absolute value) where

$$x = \pi(g_0 + d_{G\alpha})/d_{Gd} \tag{9.5}$$

for some g_0.

The displayed range of values of g will be $g \in [-3, 3]$.

9.3 Comparison of Approximate Eigenfunctions and Eigenvalues to the Tangent Representation

In this section we display the graphs of the approximate Eigenfunctions and Eigenvalue functions, and the proposed tangent graphs for SU(2), SU(3) and SU(4).

9.3.1 SU(2)

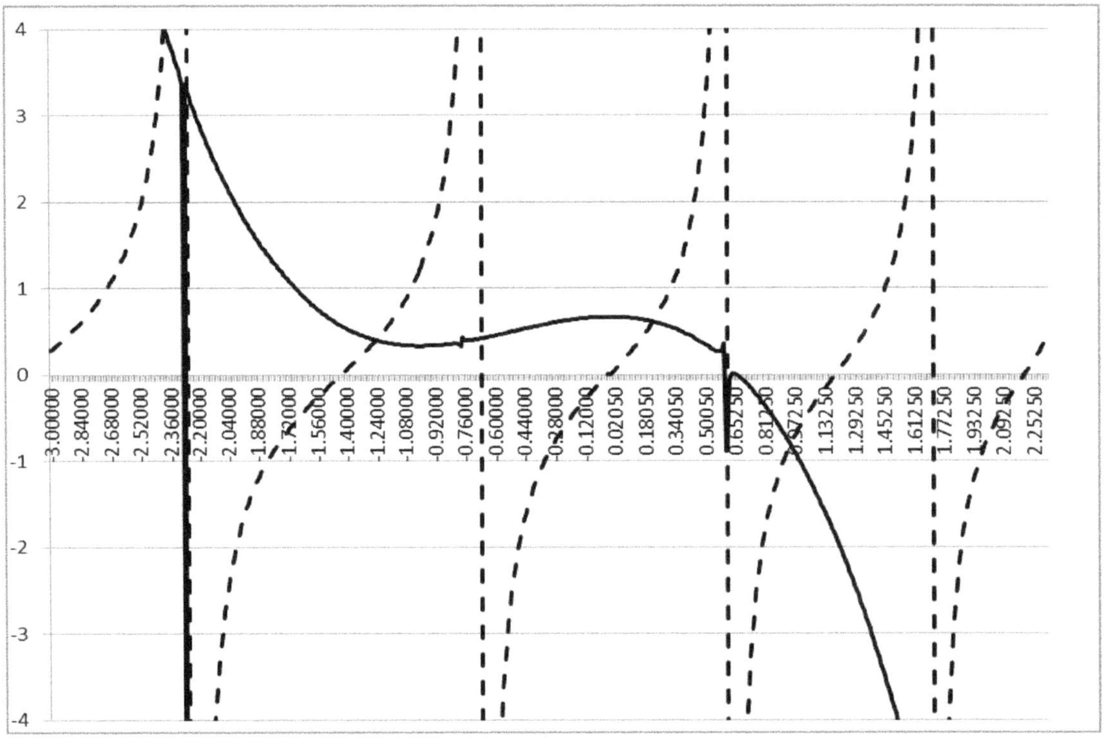

Figure 9.1 The SU(2) eigenvalue function F_1 is a solid line, while the tangent form of F_1 of eq. 9.1 is the broken line. The constants are d_{Gd} = 1.29911 - 0.08929g, and d_{Gf} = 0. F_1 is plotted vertically. The exponent g is plotted horizontally.

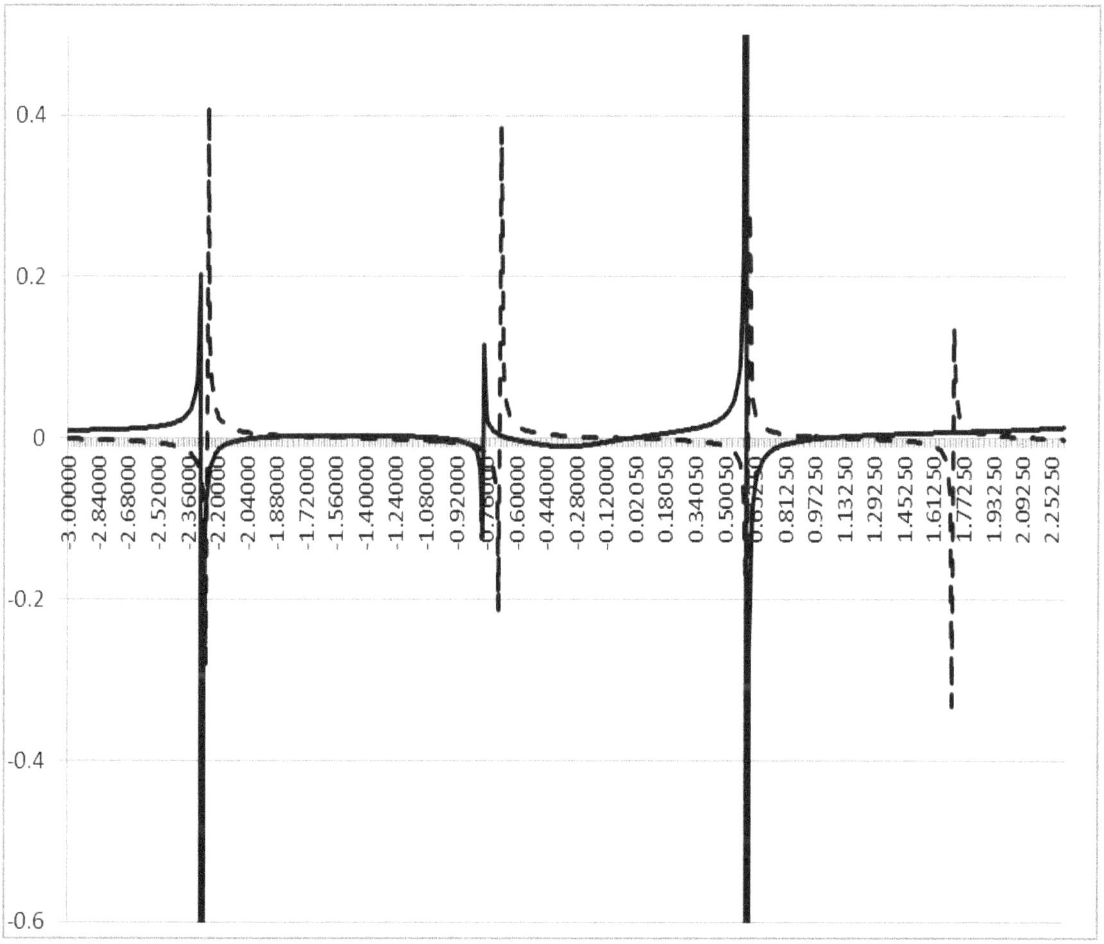

Figure 9.2 The SU(2) eigenvalue function α_G is a solid line, while the tangent form of α_G of eq. 9.2 is the broken line. The constants are d_{Gd} = 1.29911 - 0.08929g, and the α_G constant is $d_{G\alpha}$ = 0.00348417. α_G is plotted vertically. The exponent g is plotted horizontally.

9.3.2 SU(3)

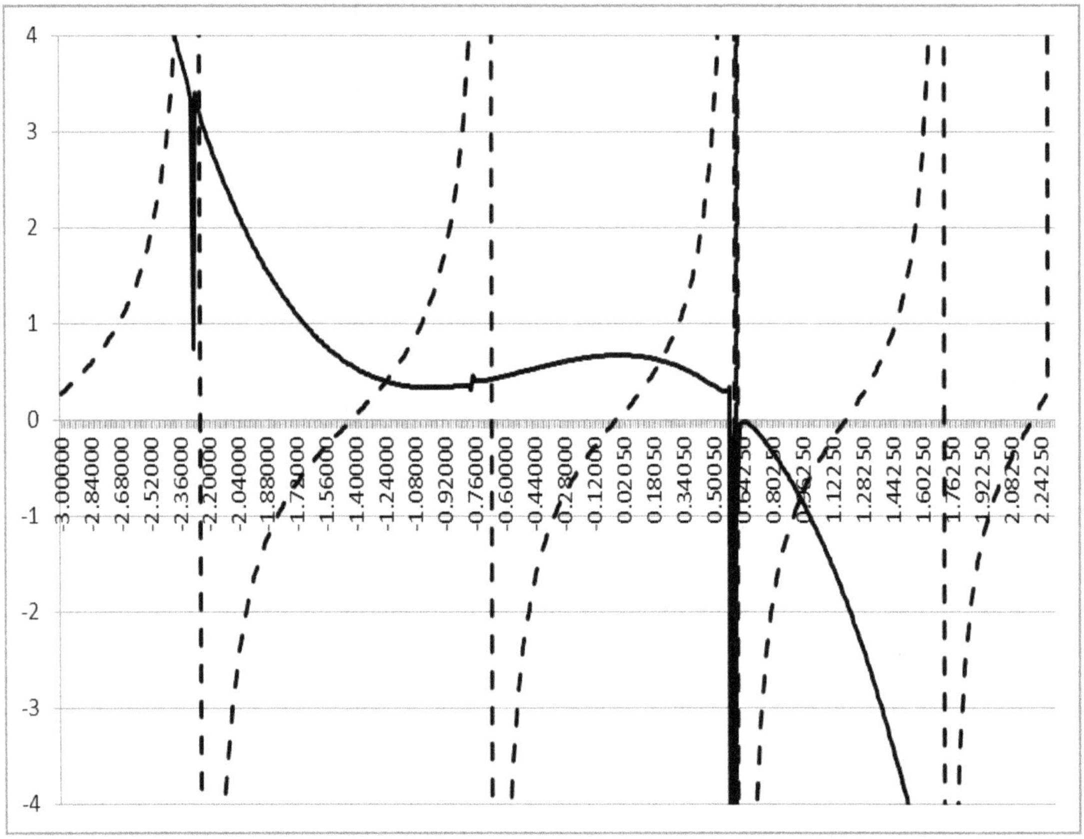

Figure 9.3 The SU(3) eigenvalue function F_1 is a solid line, while the tangent form of F_1 of eq. 9.1 is the broken line. The constants are d_{Gd} = 1.29911 - 0.08929g, and d_{Gf} = 0. F_1 is plotted vertically. The exponent g is plotted horizontally.

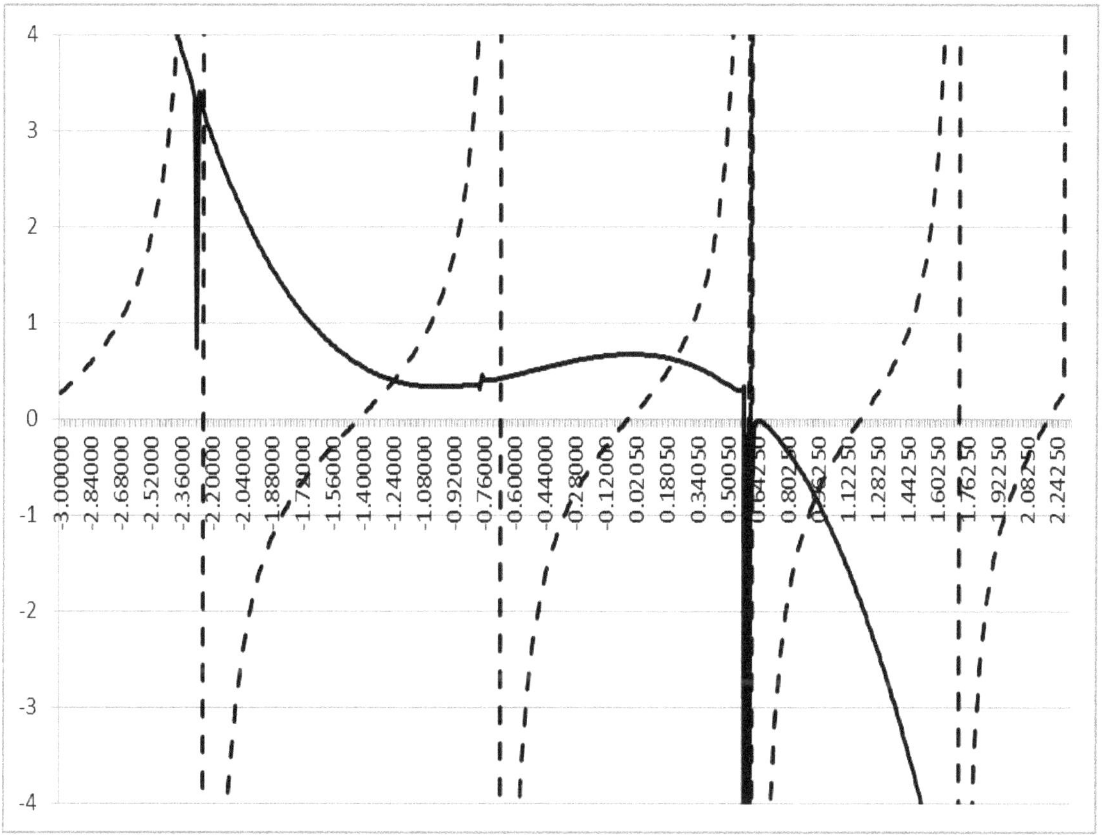

Figure 9.4 The SU(3) eigenvalue function α_G is a solid line, while the tangent form of α_G of eq. 9.2 is the broken line. The constants are d_{Gd} = 1.29911 - 0.08929g, and the α_G constant is $d_{G\alpha}$ = 0.00348417. α_G is plotted vertically. The exponent g is plotted horizontally.

9.3.3 SU(4)

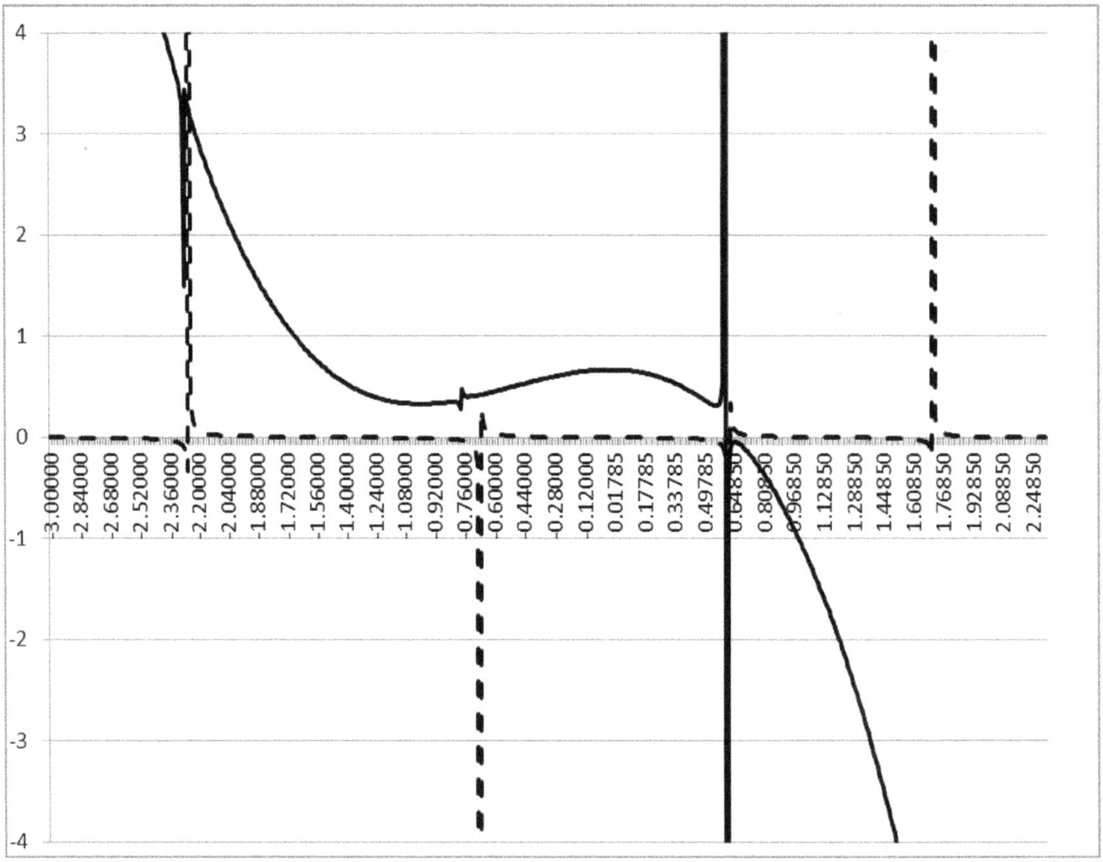

Figure 9.5 The SU(4) eigenvalue function F_1 is a solid line, while the tangent form of F_1 of eq. 9.1 is the broken line. The constants are d_{Gd} = 1.29911 - 0.08929g, and d_{Gf} = 0. F_1 is plotted vertically. The exponent g is plotted horizontally.

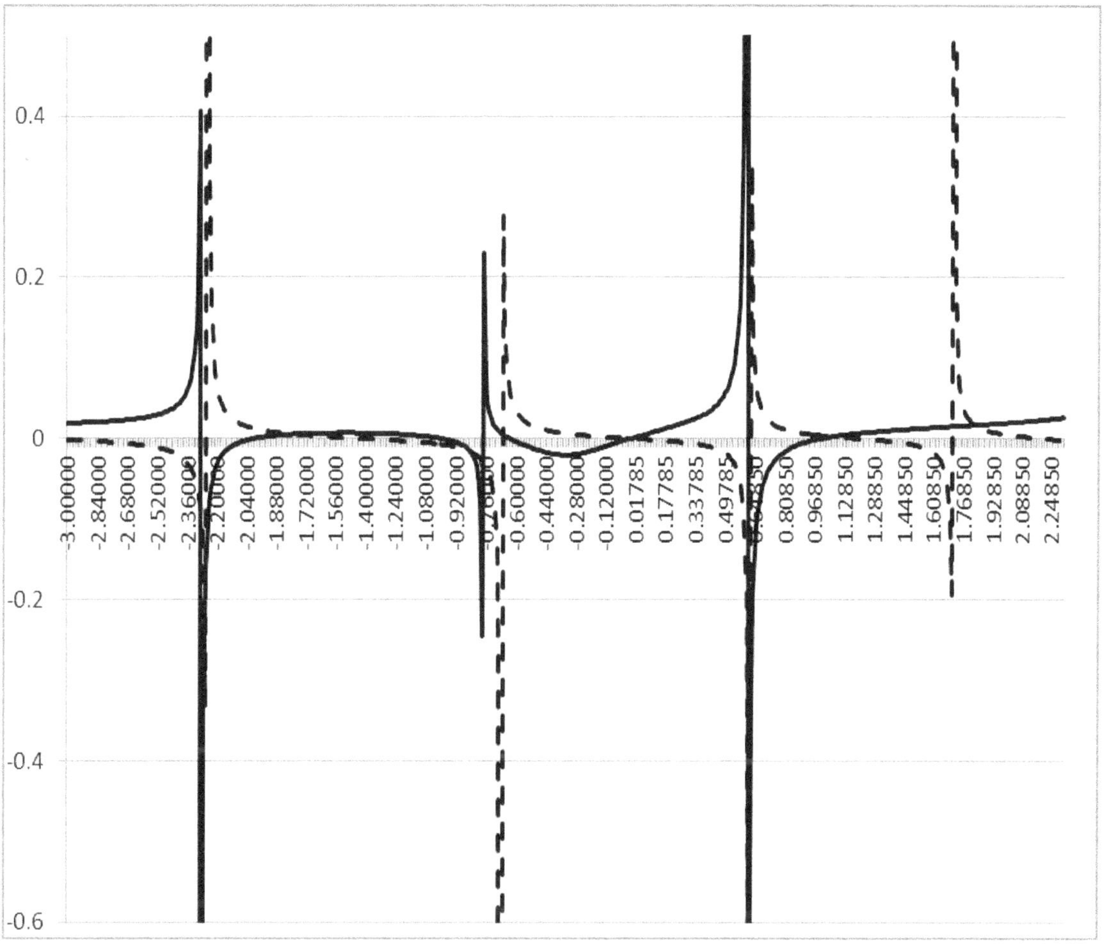

Figure 9.6 The SU(4) eigenvalue function $\alpha_G(g)$ is a solid line, while the tangent form of α_G of eq. 9.2 is the broken line. The constants are d_{Gd} = 1.29911 - 0.08929g, and the α_G constant is $d_{G\alpha}$ = 0.00348417. α_G is plotted vertically. The exponent g is plotted horizontally.

9.4 Comments

The quantities d_{Gd} and $d_{G\alpha}$ have the same values for SU(2), SU(3) and SU(4). The values are the same for QED in chapter 11 as well. The group constants c_G differ from group to group. *The similarity in form for all non-abelian cases is due to our use of a "universal" form for eigenfunctions. Our justification is our success in the approximate calculation of all coupling constants.*

The similarity in the form, and the values of c_G *and other constants in eq. 9.2, and in chapter 11, "explain" the difference in Standard Model coupling constants.*

10. Approximate Non-Abelian, Running Coupling Constants with β Functions to All Orders in Perturbation Theory

In this chapter we consider the one fermion loop approximation for the β function of the Callan-Symanzik equation:

$$-\beta_1(\alpha) + m\partial(\alpha\pi^{(1)}{}_c)/\partial m \approx 0 \qquad (10.1)$$

where the vacuum polarization part is

$$\pi^{(1)}{}_c = F_1(\alpha)\ln(-q^2/m^2) + \ldots \qquad (10.2)$$

resulting in the β function term

$$\beta_1(\alpha) = -2\alpha F_1(\alpha) \qquad (10.3)$$

The integration of β_1 yields

$$\ln(m_2/m_1) = \int_{\alpha_1}^{\alpha_2} d\alpha/\beta(\alpha) \qquad (10.4)$$

where

$$\alpha_i = \alpha(m_i) \qquad (10.5)$$

10.1 Approximate Form of the Eigenvalue Function $F_1(\alpha)$ to all Perturbative Orders

We first calculate the eigenvalue function $F_1(\alpha)$ and then proceed to calculate $\beta_1(\alpha)$. Then eq. 10.4 becomes soluble and an expression for the running coupling constant can be calculated.

For the sake of a transparent physical result[41] we assume $d_{Gd} = 1.5$, which is a reasonable approximation for g in the range [-3, 3] (the domain of physical interest).

Eq. 9.2 can be inverted to yield $g(\alpha)$ for a given group G:

$$g(\alpha) = (d_{Gd}/\pi) \arctan(\alpha_G/c_G) - d_{G\alpha} \qquad (10.6)$$

Substituting for g in eq. 9.1 using eq. 10.6 in gives

$$F_{G1}(\alpha) = [\alpha - c_G \tan(h_G)]/[1 + \alpha c_G \tan(h_G)] \qquad (10.7)$$

where

$$h_G = \pi(d_{G\alpha} - d_{Gf})/d_{Gd} \qquad (10.8)$$

Eq. 10.7 is a remarkably simple expression for the eigenvalue function in terms of α. It has the "eigenvalue zero" at[42]

$$\alpha_G = |c_G \tan(h_G)| \qquad (10.9)$$

We will see in chapter 11 that the QED[43] Fine Structure Constant α has the same form making it analogous to the Madhava-Leibniz formula for π discussed in section 9.2.

For massless QED, given the known value of $\alpha = 0.007297353$ and using $\alpha = \tan(h)$ we find

---

[41] The expression used in our plots $d_{Gd} = 1.29911 - 0.08929g$ could have been used and would lead to a tractable calculation. The essence of the physics is well brought out by approximating $d_{Gd} = 1.5$.

[4242] Eq. 10.9 uses an absolute value for α_G since coupling constants are real-valued.

[43] The use of intermediate renormalization and thus F_2 for QED does not change the character of the results. Only eq. 10.3 requires a superficial change.

$$h = 0.007297223473 \tag{10.10}$$

and the QED quantities

$$h/\pi = (d_\alpha - d_f)/d_d = 0.002322780335 \tag{10.11}$$

Using $d_\alpha = 0.00348417$ and $d_f = 0$ from chapter 11 we find

$$d_d = 1.499999784 \approx 1.5 \tag{10.12}$$

10.2 Approximate Form of Running Coupling Constant to all Perturbative Orders

Eqs. 10.3, 10.4 and 10.7 determine the approximate running coupling constant

$$\ln(m_2/m_1) = \int_{\alpha_1}^{\alpha_2} d\alpha/(-2\alpha F_{G1}(\alpha)) \tag{10.4}$$

The running coupling constant integration yields

$$(m_1/m_2)^2 = [(\alpha_2 - c_G \tan(h_G))/(\alpha_1 - c_G \tan(h_G))]^y (\alpha_2/\alpha_1)^z \tag{10.13}$$

where

$$y = c_G \tan(h_G) + (c_G \tan(h_G))^{-1} \tag{10.14}$$

$$z = -(c_G \tan(h_G))^{-1} \tag{10.15}$$

If

$$\alpha_i >> |c_G \tan(h_G)|$$

for i = 1, 2, then eq. 10.13 becomes approximately

$$(m_1/m_2)^2 = [\alpha_2/\alpha_1]^{c_G \tan(h_G)} = [\alpha_2/\alpha_1]^{\alpha_G \mathrm{signum}(c_G)} \tag{10.16}$$

where the exponent α_G is the constant coupling constant value in eq. 10.9 and signum(c_G) is the sign of c_G giving the proper asymptotic freedom of non-abelian couplings and the infinite bare charge of the QED (abelian) case.

Eq. 10.16 can be reexpressed as

$$\alpha(m_2) = \alpha_1(m_1)\,(m_1/m_2)^{2\mathrm{signum}(c_G)/\alpha_G} \qquad (10.17)$$

10.3 Running Coupling Constant

The running coupling constant expression is applicable to QED, Weak SU(2), and Strong SU(3). Eqs. 10.13 and 10.17 show QED's running coupling constant gets stronger as the mass increases, while the non-abelian running coupling constants decrease as the mass increases yielding asymptotic freedom.

The growth/decline of the running coupling constant is a power law in this formulation while in other formulations, based on low order prturbation theory, it is logarithmic.[44] This difference impacts on GUT unified theories.

---

[44] H. Georgi, H. R. Quinn, and S. Weinberg, Phys. Rev. Lett. **33**, 451 (1974); W. E. Caswell, Phys. Rev. Lett. **33**, 244 (1974).

11. Approximate Massless QED Eigenvalue Function to All Orders in Perturbation Theory

The approximate calculation of F_2 in perturbation theory to all orders in chapter 5 and Appendix A also resembles a tangent approximation with

$$d_d = d_{Gd} = 1.29911 - 0.08929*g$$
$$F_2(g) = \tan(\pi g/d_d)$$

and

$$\alpha(g) = \tan(\pi(g + 0.00348417)/d_d)$$

using the notation of chapter 9 and eqs. 10.10-10.12. Figs. 11.1 and 11.2 show the close approximation given by the tangent curves.

The general form of the expression for the fine structure constant α is

$$\alpha_G = |c_G \tan(x)| \qquad (9.4)$$

where

$$x = \pi(g_0 + d_{G\alpha})/d_{Gd} \qquad (9.5)$$

for some g_0.

The QED α value is

$$\alpha = 0.007297353$$
$$\alpha^{-1} = 137.0359991$$

We have shown the general form for α is

$$\alpha = \tan(h) \qquad (11.1)$$

in close analogy to the Madhava-Leibniz representation of $\pi = 3.14159\ldots$.

We use

$$h = 0.007297223473 \quad (10.10)$$

in eq. 11.1 based on the α calculation of chapter 5, and the QED quantities

$$h/\pi = (d_\alpha - d_f)/d_d = 0.002322780335 \quad (10.11)$$

Thus the fine structure constant α is a feature of the dynamics of QED and is not a cosmological result, or a result of any other theory.

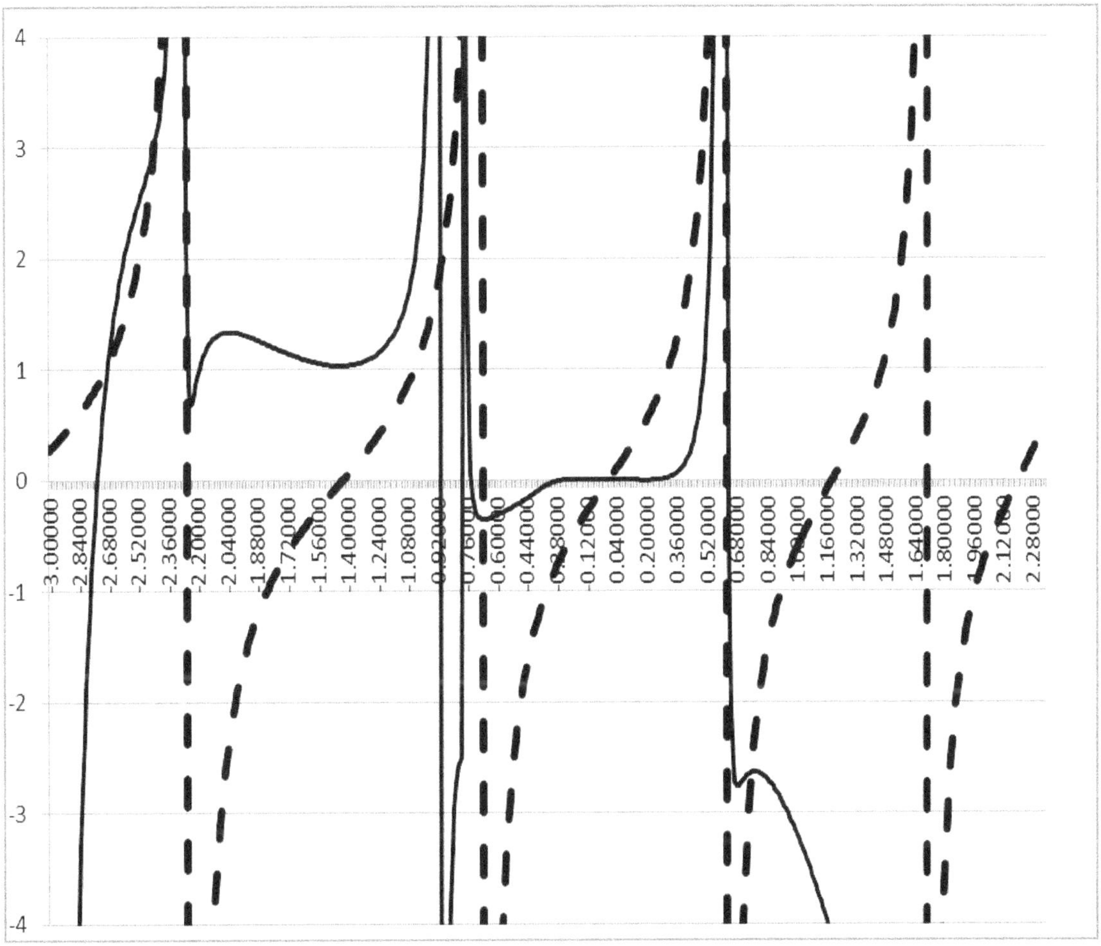

Figure 11.1 The massless QED eigenvalue function F_2 is the solid line, while the tangent form of F_2 of eq. 9.1 is the broken line. The constants are d_{Gd} = 1.29911 - 0.08929g, and d_{Gf} = 0. F_2 is plotted vertically. The exponent g is plotted horizontally.

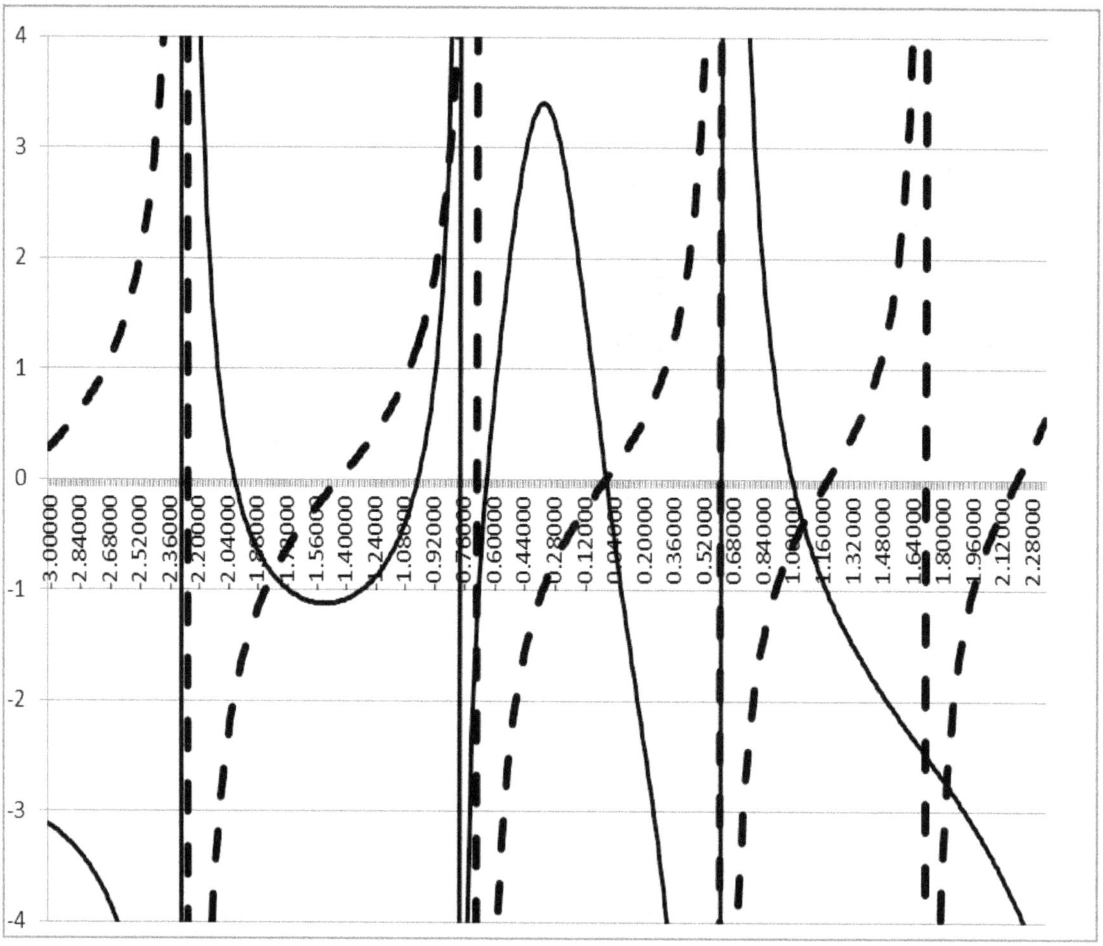

Figure 11.2 The QED fine structure eigenvalue function α(g) is a solid line, while the tangent form of α(g) of eq. 9.2 is the broken line. The constants are d_{Gd} = 1.29911 - 0.08929g, and the α_G constant is $d_{G\alpha}$ = 0.00348417 for G = QED. α(g) is plotted vertically. The exponent g is plotted horizontally.

Appendix A. "Approximate Calculation of the Eigenvalue Function in Massless Quantum Electrodynamics"

This refereed paper is S. Blaha, Phys. Rev. **D9**, 2246 (1974). Reprinted with the kind permission of Physical Review D.

PHYSICAL REVIEW D VOLUME 9, NUMBER 8 15 APRIL 1974

Approximate calculation of the eigenvalue function in massless quantum electrodynamics*

Stephen Blaha†
Department of Physics, University of Washington, Seattle, Washington 98195
(Received 24 August 1973)

We solve a vertex equation in massless quantum electrodynamics and use the results to calculate an approximation to the eigenvalue function, F_1, in the Johnson-Baker-Willey model. This approximation consists of a summation of the contributions to F_1 of all one-electron-loop diagrams in which no internal photon lines intersect (if all such lines are drawn within the electron loop). Our result reproduces the known low-order terms, $F_1 = 2/3 + \alpha/2\pi - \frac{1}{4}(\alpha/2\pi)^2$ exactly. In addition we find branch-point singularities and zeros at points corresponding to values for the fine-structure constant of order unity. Nonperturbative solutions of the vertex equation and F_1 are also shown to exist.

I. INTRODUCTION

The apparently divergent renormalization constants in quantum electrodynamics (QED) have been a source of uneasiness for many years. Some have taken this property to reflect a fundamental incompleteness of QED. Johnson, Baker, and Willey[1] have responded to the problem by developing a model QED which has finite renormalization constants if a certain function, denoted F_1, of the fine-structure constant is zero. $F_1(\alpha)$ is the coefficient of $(\alpha/2\pi)\ln\Lambda^2$ in the sum of the contributions of all one-electron-loop vacuum-polarization diagrams to Z_3^{-1}. [Examples of $O(\alpha^2)$ and $O(\alpha^3)$ diagrams contributing to F_1 are given in Figs. 1 and 2.]

Adler[2] enhanced interest in F_1 by noting that any zero of F_1 must be of infinite order (an essential singularity) and by raising the possibility that the zero might occur at the value of the physical fine-structure constant. In this case, the requirement that QED be finite would determine the fine-structure constant, and QED would then be a self-contained theory (but for the choice of the electron mass scale).

Because of the importance of this possibility, a calculation of F_1 would be of great interest. Since an exact calculation appears unlikely at the moment, we have calculated an approximate expression for F_1 based on an extrapolation of a property of its low-order terms. In low order[3]

$$F_1 = \frac{2}{3} + \frac{\alpha}{2\pi} - \frac{1}{4}\left(\frac{\alpha}{2\pi}\right)^2 + \cdots . \tag{1}$$

The third term was first calculated by Rosner in the Landau gauge. Brandt[4] recalculated this term in the Feynman gauge and found that the contribution to F_1 of a certain subset of the diagrams summed to zero [see Fig. 1(a)]. These same diagrams contained the only appearances of the zeta function, $\zeta(3)$. The diagrams of Fig. 1(b) sum to the total $O(\alpha^2)$ term of F_1. [Due to Brandt's renormalization procedure, each electromagnetic vertex in the contributing set of Fig. 1(b) should be understood to represent a zero-momentum-transfer vertex function rather than a Dirac matrix.] If Brandt's result generalizes to higher order, one could hope to find F_1 by summing a much simpler subset of diagrams. The obvious generalization is to sum only those diagrams in which no internal photon lines intersect when all photon lines are drawn within the electron loop. Figure 2 shows these diagrams in $O(\alpha^3)$.

By choosing to sum only a special subset of the diagrams, we have lost gauge invariance. However, there are two situations where this might be an acceptable loss: (1) where the sum of the selected subset of diagrams indeed reproduces the exact gauge-invariant result (making the question academic), and (2) where the sum of the diagrams gives the dominant contribution to F_1 so that one can observe the important features, e.g., an essential singularity.

We will calculate the contributions of our selected subset of diagrams to F_1 by finding the vertex function and electron propagator which generate the contributions of these diagrams when substituted in the photon self-energy. The gauge in which we work will be determined by requiring that the vertex function and electron propagator (for the set of diagrams considered) satisfy the differential Ward identity.

In Sec. II we derive the necessary vertex equation. Section III gives the details of the calculation. Section IV contains a discussion of the important features of our results.

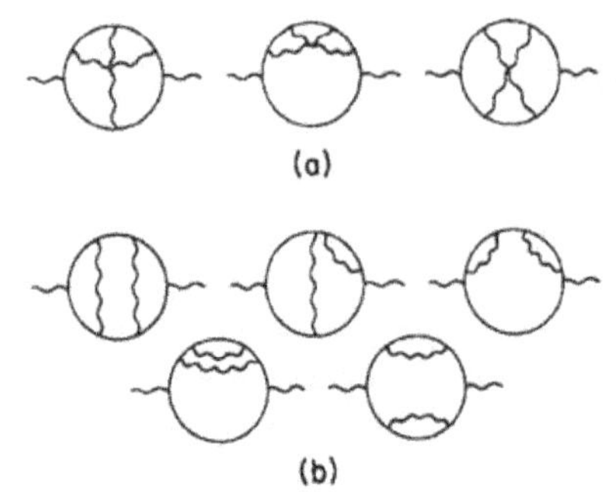

FIG. 1. (a) Diagrams whose contribution to F_1 in the Feynman gauge sums to zero. (b) Diagrams which give the total contribution to F_1 in $O(\alpha^2)$ in the Feynman gauge.

II. DERIVATION OF THE VERTEX EQUATION

Before deriving the vertex equation, we will define the relation of F_1 to the vertex function. Johnson, Willey, and Baker[5] relate F_1 and the contribution of one-electron-loop diagrams to $\Pi_{\mu\nu}$, the photon self-energy:

$$\Pi_{\mu\mu\beta\beta} \equiv \frac{\partial^2}{\partial q^\beta \partial q_\beta} \Pi^\mu_\mu(q)\bigg|_{q=0}$$
$$= \frac{-12\alpha}{\pi} F_1 \int \frac{dp^2}{p^2} + \cdots, \tag{2}$$

where q is the external photon momentum. They then show

$$\Pi_{\mu\mu\beta\beta} = \frac{-ie^2}{(2\pi)^4} \int d^4p \, \mathrm{Tr}(\Gamma_\mu G^\beta_\beta \Gamma^\mu + 2\Gamma_{\mu\beta} G^\beta \Gamma^\mu + \Gamma_\mu G K^\beta_\beta G \Gamma^\mu + 2\Gamma_\mu G_\beta K^\beta G \Gamma^\mu + 2\Gamma^\mu G K^\beta G \Gamma_{\mu\beta}), \tag{3}$$

where G signifies the introduction of the unrenormalized electron propagator S_F at appropriate places in the traces, $S_F \cdots S_F$; K is the electron-positron scattering kernel; Γ_μ is the vertex function; and the appearance of the subscript (superscript) β on a quantity indicates that a derivative is to be taken with respect to the external photon momentum, q, before setting $q=0$. In the case we will consider below K is independent of q; ($K_\beta = K^\beta_\beta = 0$), and the explicit form $\Pi_{\mu\mu\beta\beta}$ assumes is

$$\Pi_{\mu\mu\beta\beta} = \frac{-ie^2}{(2\pi)^4} \mathrm{Tr} \int d^4p (\Gamma_\mu S_F \Gamma^\mu S^\beta_{F\beta} + \Gamma_\mu S^\beta_{F\beta} \Gamma^\mu S_F + 2\Gamma^\mu S_{F\beta} \Gamma_\mu S^\beta_F + 2\Gamma_\mu S^\beta_F \Gamma^\mu_\beta S_F + 2\Gamma^\mu S_F \Gamma_{\mu\beta} S^\beta_F). \tag{4}$$

The quantities we must find are $\Gamma_\mu(p,p)$, the zero-momentum-transfer vertex function, S_F, and $\Gamma_{\mu\beta} = (\partial/\partial q^\beta)\Gamma_\mu|_{q=0}$. We will find expressions for these quantities which generate the contributions to F_1 of the diagrams described in the Introduction.

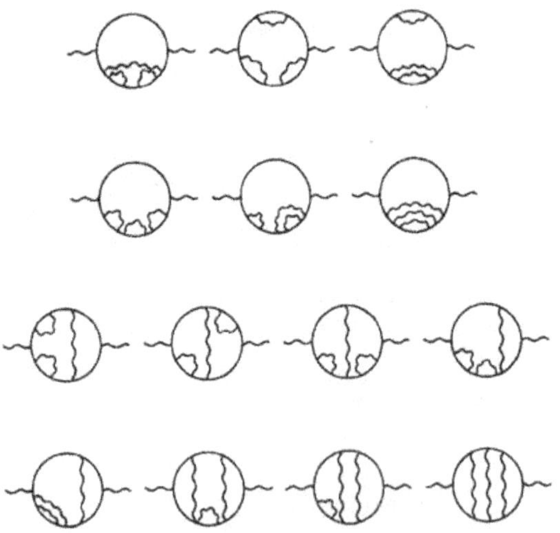

FIG. 2. The distinct diagrams contributing to the $O(\alpha^3)$ term in F_1 in our approximation.

Our starting point will be the equation for the vertex function in massless QED with no self-energy insertions in the photon propagator:

$$\Gamma_\mu(p_+,p_-) = \gamma_\mu + \frac{ie^2}{(2\pi)^4} \int d^4k \, \Gamma_\lambda(p_+,k_+) S_F(k_+) \Gamma_\mu(k_+,k_-) \times S_F(k_-) \Gamma_\sigma(k_-,p_-) D^{\lambda\sigma}(p-k) + \cdots, \tag{5}$$

where $p_\pm = p \pm \frac{1}{2}q$ are the external electron momenta, $k_\pm = k \pm \frac{1}{2}q$, and $D_{\lambda\sigma}$ is the photon propagator.

Since we are only interested in diagrams with nonintersecting lines, we need only keep the exhibited terms on the right-hand side of the equation. In addition, if we kept strictly to nonintersecting photon line diagrams, we would substitute γ_ν for Γ_ν and γ_σ for Γ_σ in the second term. But the suggestion of Brandt's work was to place zero-momentum-transfer vertex functions at all photon vertices. We choose to do this in the following manner:

$$\Gamma_\mu(p_+,p_-) = \gamma_\mu + \frac{ie^2}{(2\pi)^4} \int d^4k \, \Gamma_\lambda(k,k) S_F(k_+) \Gamma_\mu(k_+,k_-) \times S_F(k_-) \Gamma_\sigma(k,k) D^{\lambda\sigma}(p-k), \tag{6}$$

where we have introduced zero-transfer vertices which are only functions of the loop integration variable. We will discuss the results of other possible choices such as $\Gamma_\nu(p,p) \cdots \Gamma_\sigma(p,p)$ in Sec. IV. Since $\Gamma_\nu(p_+,k_+)$ and $\Gamma_\sigma(k_-,p_-)$ can be expanded in a double power series whose first terms are $\Gamma_\nu(k,k)$ and $\Gamma_\sigma(k,k)$, respectively, it is clear that we are not introducing any spurious contributions by our choice at these vertices. Figure 3 gives a graphical representation of Eq. (6).

Having chosen our vertex equation, we now complete our set of equations by requiring that the Ward identity be satisfied,

$$\Gamma_\mu = \frac{\partial}{\partial p^\mu} S_F^{-1}(p). \tag{7}$$

Equations (6) and (7) can only be simultaneously satisfied in a special gauge. This gauge is the Feynman gauge up to terms of $O(\alpha)$. In a preliminary study[6] of these equations, the gauge function was set equal to zero (which was reasonable in an investigation of the region near $\alpha=0$) and only the component of the vertex equation kept which could be obtained by multiplying Eq. (6) by γ_μ and summing over μ. The results of that investigation will be seen to be in qualitative agreement with these results. To find F_1 we must solve two equations obtained from Eq. (6) by setting $q=0$ in the first case, and setting $q=0$ after taking a derivative in the second:

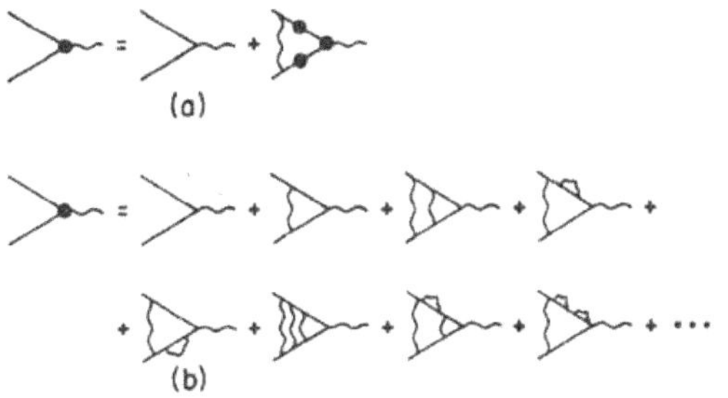

FIG. 3. (a) Diagrammatic representation of the vertex equation. (b) Some low-order vertex diagrams embodied in the vertex equation. All seemingly bare vertices (except the vertex to which the external photon couples) actually represent zero-momentum-transfer vertex functions.

$$\Gamma_\mu(p) = \gamma_\mu + \frac{ie^2}{(2\pi)^4}\int d^4k\, \Gamma_\lambda(k)S_F(k)\Gamma_\mu(k)S_F(k)\Gamma_\sigma(k)D^{\lambda\sigma}(p-k), \tag{8}$$

$$\Gamma_{\mu\alpha} = \frac{ie^2}{(2\pi)^4}\int d^4k\, D^{\lambda\sigma}(p-k)\Big[\Gamma_\lambda(k)S_F(k)\Gamma_{\mu\alpha}(k)S_F(k)\Gamma_\sigma(k) + \Gamma_\lambda(k)\Big(\frac{\partial}{\partial k^\alpha}S_F\Big)\Gamma_\mu(k)S_F(k)\Gamma_\sigma(k) - \Gamma_\lambda(k)S_F(k)\Gamma_\mu(k)\Big(\frac{\partial}{\partial k^\alpha}S_F(k)\Big)\Gamma_\sigma(k)\Big], \tag{9}$$

where

$$\Gamma_\mu(k) = \Gamma_\mu(k,k)$$

and

$$\Gamma_{\mu\alpha}(p) = 2\frac{\partial}{\partial q^\alpha}\Gamma_\mu(p_+,p_-)\Big|_{q=0}, \tag{10}$$

with $p_\pm = p \pm \frac{1}{2}q$. [Note that $\Gamma_{\mu\beta}$ in Eq. (10) is a factor of 2 different from the similar-looking quantity in Eqs. (3) and (4). All subsequent appearances of $\Gamma_{\mu\beta}$ conform to Eq. (10).] The photon propagator has the form

$$D_{\mu\nu}(k) = -\Big(\frac{g_{\mu\nu}}{k^2} - \psi\frac{k_\mu k_\nu}{k^4}\Big), \tag{11}$$

where $\psi=\psi(\alpha)$ will be determined by requiring Eqs. (7) and (8) to be consistent. In the following we will perform a Wick rotation on all four-vectors in Eqs. (7), (8), and (9), and introduce an ultraviolet cutoff, Λ, in all radial integrations.

One approach to solving our equations would be to iterate Eq. (8) and calculate Γ_μ and S_F order by order in α. The first iteration of Eq. (8) gives the triangle diagram which contributes surface terms to Γ_μ which are proportional to p^2/Λ^2. For terms of this sort we follow the procedure of Gell-Mann and Low[7] (Sec. III) and "drop terms that approach zero as $\Lambda^2 \to \infty$." Order by order in α such terms should be dropped as they appear. As a result

$$S_F^{-1} = \not p B(p^2) = \sum_{i,j=0}^{\infty} c_{ij}\,\alpha^i[\ln(p^2/\Lambda^2)]^j,$$

with c_{ij} a constant satisfying $c_{ij}=0$ if $j > j_0(i)$; i.e., in any order of α there is an upper bound on the power of divergent logarithms. Equations (6) and (7) allow us to calculate c_{ij} recursively (dropping terms violating the criteria on c_{ij} such as p^2/Λ^2).

Actually, since the terms we are dropping are surface terms we may find the form of the "kept" terms of B by converting Eq. (6) into a differential equation after performing angular integrations. The surface terms are now eliminated and scaling arguments show $B = f(p/\Lambda)^{2g}$ with f and g constants. This is *not* a solution of Eq. (6), because of surface terms. However, substitution of this solution into the right-hand side of Eq. (6) and expansion of terms in double power series

$$\sum d_{ij}\,\alpha^i[\ln(p^2/\Lambda^2)]^j$$

will allow us to uniquely determine the surface terms which would have been dropped in a recursive solution. In the case at hand we drop terms proportional to p^2/Λ^2 so that equating coefficients on either side of the equation will not lead to a violation of $c_{ij}=0$ for $j>j_0(i)$ in B's expansion.

As a result of the form of B we have

$$\Gamma_\mu(p)=f\left(\gamma_\mu+2g\frac{\not{p}p_\mu}{p^2}\right)\left(\frac{p}{\Lambda}\right)^{2\epsilon}, \tag{12}$$

$$S_F=\left[f\not{p}\left(\frac{p}{\Lambda}\right)^{2\epsilon}\right]^{-1}, \tag{13}$$

$$\Gamma_{\mu\alpha}=\frac{f_3}{p^2}(\not{p}\gamma_\mu\gamma_\alpha-\gamma_\alpha\gamma_\mu\not{p})\left(\frac{p}{\Lambda}\right)^{2\epsilon}, \tag{14}$$

which implies

$$F_1=\tfrac{2}{3}(1-\tfrac{3}{2}g^2-g^3)-f_3/f\,. \tag{15}$$

Equation (9) determines f_3, which is the product of a function of g and ψ times f. Thus f_1 depends only on g and ψ [cf. Eq. (58)] and surface terms of the kind mentioned above do not affect F_1.

Rather than calculate F_1 using the approach just outlined, we will substitute Eqs. (12), (13), and (14) into Eqs. (8) and (9) and determine g, ψ, f, and f_3 in this more economical way.

III. THE CALCULATION

In this section we solve Eqs. (7), (8), and (9) by showing that Eqs. (12), (13), and (14) are consistent solutions of the equations for appropriate choices of f, g, and f_3.

We begin by substituting Eqs. (12) and (13) into Eq. (8). In order to illustrate our approach, we will now use a fact which will be apparent at the end; namely, that Eq. (8) has the following form after performing the loop integral:

$$f\left(\gamma_\mu+2g\frac{p_\mu\not{p}}{p^2}\right)\left(\frac{p}{\Lambda}\right)^{2\epsilon}=\gamma_\mu+A\gamma_\mu+B\gamma_\mu\left(\frac{p}{\Lambda}\right)^{2\epsilon}+C\frac{\not{p}p_\mu}{p^2}\left(\frac{p}{\Lambda}\right)^{2\epsilon}+D\frac{\not{p}p_\mu}{p^2}, \tag{16}$$

where A, B, C, D are constants. As a result

$$f=B, \tag{17}$$

$$2gf=C, \tag{18}$$

$$1+A=0, \tag{19}$$

$$D=0 \tag{20}$$

are required for consistency. We will begin by calculating the two relations resulting from multiplying Eq. (16) by γ_μ and summing over μ:

$$f(4+2g)=4B+C \tag{21}$$

and

$$0=1+A, \tag{22}$$

assuming $D=0$. Later we will calculate C and show $D=0$.

Initially Eq. (8) has the form

$$f(4+2g)\left(\frac{p}{\Lambda}\right)^{2\epsilon}=4+\frac{ie^2f}{4(2\pi)^4}\int\frac{d^4k}{k^4}D^{\lambda\sigma}(p-k)\left(\frac{k}{\Lambda}\right)^{2\epsilon}\mathrm{Tr}\gamma^\mu(\gamma_\lambda\not{k}+2gk_\lambda)\left(\gamma_\mu+2g\frac{\not{k}k_\mu}{k^2}\right)(\not{k}\gamma_\sigma+2gk_\sigma) \tag{23}$$

after taking the trace of γ_μ times Eq. (8).

The γ-matrix algebra may be performed in the integrand, and one arrives at the following expression:

$$f(4+2g)\left(\frac{p}{\Lambda}\right)^{2\epsilon}=4-\frac{ie^2f}{(2\pi)^4}\int\frac{d^4k\,k^{2\epsilon-2}}{(p-k)^2\Lambda^{2\epsilon}}\left(4w-\frac{\psi}{(p-k)^2k^2}\{4v[k\cdot(p-k)]^2-2gk^2(p-k)^2\}\right) \tag{24}$$

or

$$f(4+2g)\left(\frac{p}{\Lambda}\right)^{2\epsilon}=4-\frac{ie^2f}{(2\pi)^4}[(4w+2g\psi)I_1-\psi v(I_2+2I_3+I_4)], \tag{25}$$

where $w=1+3g+6g^2+2g^3$, $v=1+5g+6g^2+2g^3$, and

$$I_1=\int\frac{d^4k}{(p-k)^2k^2}\left(\frac{k}{\Lambda}\right)^{2\epsilon}\equiv\frac{i\pi^2}{g}\left[1-\frac{1}{g+1}\left(\frac{p}{\Lambda}\right)^{2\epsilon}\right], \tag{26}$$

$$I_2=\int\frac{d^4k}{k^4}\left(\frac{k}{\Lambda}\right)^{2\epsilon}\equiv\frac{i\pi^2}{g}, \tag{27}$$

$$I_3=\int\frac{d^4k(k^2-p^2)}{(p-k)^2k^4}\left(\frac{k}{\Lambda}\right)^{2\epsilon}\equiv\frac{i\pi^2}{g}\left[1+\frac{2}{g^2-1}\left(\frac{p}{\Lambda}\right)^{2\epsilon}\right], \tag{28}$$

$$I_4=\int\frac{d^4k(k^2-p^2)^2}{(p-k)^4k^4}\left(\frac{k}{\Lambda}\right)^{2\epsilon}\equiv\frac{i\pi^2}{g}\left[1+\frac{2g}{g^2-1}\left(\frac{p}{\Lambda}\right)^{2\epsilon}\right]. \tag{29}$$

We have assumed $g>0$ and used

$$4[k\cdot(p-k)]^2=(p^2-k^2)^2-2(p^2-k^2)(p-k)^2+(p-k)^4\,.$$

The method of evaluating the integrals given above is described in the Appendix. Substituting for the integrals in Eq. (25), we obtain the following two equations:

$$4+2g=\frac{-\alpha}{4\pi}\left[\frac{4w}{g(g+1)}+\frac{2\psi}{g+1}+\frac{2\psi(g+2)v}{g(g^2-1)}\right], \tag{30}$$

$$0=4g+\frac{\alpha}{\pi}fw+\frac{\alpha}{2\pi}gf\psi-\frac{\alpha}{\pi}\psi fv. \tag{31}$$

Equations (30) and (31) correspond to Eqs. (21) and (22). We now proceed to calculate C and show $D=0$. This allows us to simplify the integrand of Eq. (8) by neglecting all terms proportional to γ_μ. Terms appearing on the right-hand side of Eqs. (32)–(39) are to be taken modulo quantities proportional to γ_μ. Thus

$$2gf\frac{\not p p_\mu}{p^2}\left(\frac{p}{\Lambda}\right)^{2\epsilon}\equiv\frac{-ie^2f}{(2\pi)^4}\int\frac{d^4k}{k^2(p-k)^2}\left(\frac{k}{\Lambda}\right)^{2\epsilon}\left[\frac{4\not k k_\mu}{k^2}(1+g)(2g^2-1)-\frac{\psi}{(p-k)^2k^2}\Big(2(k_\mu p^2-p_\mu k^2)(\not p-\not k)-2\not p k_\mu(p-k)^2\right.$$
$$+2gk_\mu(p^2-k^2)(\not p-\not k)-2gk_\mu(p-k)^2\not p$$
$$\left.+8(g^2+g^3)\frac{k_\mu\not k}{k^2}[k\cdot(p-k)]^2+8g^2k\cdot(p-k)\frac{k_\mu}{k^2}(k^2\not p-k\cdot p\not k)\Big)\right] \tag{32}$$

or

$$2g\frac{\not p p_\mu}{p^2}\left(\frac{p}{\Lambda}\right)^{2\epsilon}=\frac{-ie^2}{(2\pi)^4}[4(1+g)(2g^2-1)J_{0\mu}-2\psi J_{1\mu}+2\not p\psi J_{2\mu}-2g\psi J_{3\mu}+2g\psi\not p J_{2\mu}-8g^2(g+1)J_{4\mu}-8g^2\psi J_{5\mu}], \tag{33}$$

where

$$J_{0\mu}=\int\frac{d^4k\,k_\mu\not k}{k^4(p-k)^2}\left(\frac{k}{\Lambda}\right)^{2\epsilon}\equiv\frac{-i\pi^2p_\mu\not p}{p^2(g+2)(g-1)}\left(\frac{p}{\Lambda}\right)^{2\epsilon}, \tag{34}$$

$$J_{1\mu}=\int\frac{d^4k(\not p-\not k)}{(p-k)^4k^4}(p^2k_\mu-p_\mu k^2)\equiv 0, \tag{35}$$

$$J_{2\mu}=\int\frac{d^4k\,k_\mu}{k^4(p-k)^2}\left(\frac{k}{\Lambda}\right)^{2\epsilon}\equiv\frac{-i\pi^2p_\mu}{(g^2-1)p^2}\left(\frac{p}{\Lambda}\right)^{2\epsilon}, \tag{36}$$

$$J_{3\mu}=\int\frac{d^4k\,k_\mu(p^2-k^2)(\not p-\not k)}{(p-k)^4k^4}\left(\frac{k}{\Lambda}\right)^{2\epsilon}\equiv\frac{i\pi^2p_\mu\not p}{(g+1)(g+2)p^2}\left(\frac{p}{\Lambda}\right)^{2\epsilon}, \tag{37}$$

$$J_{4\mu}=\int\frac{d^4k\,k_\mu\not k[k\cdot(p-k)]^2}{k^6(p-k)^4}\left(\frac{k}{\Lambda}\right)^{2\epsilon}\equiv\frac{i\pi^2(g^2+2g+2)p_\mu\not p}{2(g^2-1)(g^2-4)p^2}\left(\frac{p}{\Lambda}\right)^{2\epsilon}, \tag{38}$$

$$J_{5\mu}=\int\frac{d^4k\,k_\mu k\cdot(p-k)(k^2\not p-k\cdot p\not k)}{(p-k)^4k^6}\left(\frac{k}{\Lambda}\right)^{2\epsilon}\equiv\frac{-i\pi^2(2g+1)p_\mu\not p}{(g^2-1)(g^2-4)p^2}\left(\frac{p}{\Lambda}\right)^{2\epsilon}. \tag{39}$$

The evaluation of these integrals is discussed in the Appendix.
Substitution of the above expressions into Eq. (33) gives

$$g=\frac{\alpha(g+1)(1-2g^2)}{2\pi(g+2)(g-1)}-\frac{\alpha}{4\pi}\psi\frac{[(2g^3+2g^2+g-2)(g^2+2g+2)-8g^2(2g+1)]}{(g^2-1)(g^2-4)} \tag{40}$$

after some algebra. In addition it is clear that $D=0$ by examination of Eqs. (34)–(39).

Equations (30), (31), and (40) should now be solved to find $f(\alpha)$, $g(\alpha)$, and $\psi(\alpha)$. However, it is obvious that such an approach would not lead to intelligible results. Therefore, *we will parametrize* α, ψ, *and* f *with* g. We easily obtain

$$\frac{\alpha}{2\pi}=\frac{gA_4-(4+2g)A_2}{A_4A_1-A_2A_3}, \tag{41}$$

where

$$A_1=\frac{(g+1)(1-2g^2)}{(g+2)(g-1)}, \tag{42}$$

$$A_2=\frac{-(2g^3+2g^2+g-2)(g^2+2g+2)+8g^2(2g+1)}{2(g^2-1)(g^2-4)}, \tag{43}$$

$$A_3=\frac{-2(1+3g+6g^2+2g^3)}{g(g+1)}, \tag{44}$$

$$A_4=\frac{-(g+2)(1+5g+6g^2+2g^3)}{g(g^2-1)}-\frac{1}{g+1}. \tag{45}$$

In addition

$$\psi=\frac{gA_3-(4+2g)A_1}{(4+2g)A_2-gA_4} \tag{46}$$

and

$$f=\frac{-8\pi g}{\alpha[2+6g+12g^2+4g^3-\psi(2+9g+12g^2+4g^3)]}. \quad (47)$$

In Sec. IV we will discuss the properties of the above expressions and investigate the behavior of the above quantities in low order of α.

We now proceed to solve Eq. (9) for $\Gamma_{\mu\alpha}$. Substitution of Eqs. (12), (13), and (14) into Eq. (9) leads to

$$\Gamma_{\mu\alpha}(p)=\frac{-ie^2}{(2\pi)^4}\int\frac{d^4k(f_3-f)}{k^4(p-k)^2}\left(\frac{k}{\Lambda}\right)^{2\epsilon}$$
$$\times\Big[(2+4g+4g^2)(\not k\gamma_\mu\gamma_\alpha-\gamma_\alpha\gamma_\mu\not k)$$
$$-\frac{\psi}{k^2(p-k)^2}\big(\{k^4+k^2p^2-4gk^2k\cdot(p-k)+4g^2[k\cdot(p-k)]^2\}(\not k\gamma_\mu\gamma_\alpha-\gamma_\alpha\gamma_\mu\not k)$$
$$+[-k^2+2gk\cdot(p-k)][\not p\not k(\not k\gamma_\mu\gamma_\alpha-\gamma_\alpha\gamma_\mu\not k)+(\not k\gamma_\mu\gamma_\alpha-\gamma_\alpha\gamma_\mu\not k)\not k\not p]\big)\Big]. \quad (48)$$

We multiply Eq. (48) by $\not p\gamma_\alpha\gamma_\mu$ (summing over α and μ) and take the trace:

$$12f_3\left(\frac{p}{\Lambda}\right)^{2\epsilon}=\frac{-ie^2}{(2\pi)^4}(f_3-f)\int\frac{d^4k}{k^4(p-k)^2}\left(\frac{k}{\Lambda}\right)^{2\epsilon}\Big[12p\cdot k(2+4g+4g^2)$$
$$-\frac{\psi}{k^2(p-k)^2}\big(\{k^4+k^2p^2-4gk^2k\cdot(p-k)+4g^2[k\cdot(p-k)]^2\}12p\cdot k$$
$$+[-k^2+2gk\cdot(p-k)](16p\cdot k^2+8p^2k^2)\big)\Big]. \quad (49)$$

This can be rearranged into the following useful form:

$$\left(\frac{p}{\Lambda}\right)^{2\epsilon}f_3=\frac{-ie^2(f_3-f)}{12(2\pi)^4}[12(2+4g+4g^2)p_\mu J_2^\mu-\psi R_1-2g\psi R_2-48g^2\psi R_3], \quad (50)$$

where $J_{2\mu}$ was given above and

$$R_1=\int\frac{d^4k(k/\Lambda)^{2\epsilon}}{(p-k)^4k^6}[2k^2(k^2-p^2)+2k^2(p^2+k^2)(p-k)^2-4k^2(p-k)^4]=0, \quad (51)$$

$$R_2=\int\frac{d^4k(k/\Lambda)^{2\epsilon}}{k^6(p-k)^4}[4(p^2-k^2)(p^2+2k^2)-4(2p^2-k^2)(p-k)^2+4(p-k)^4]k\cdot(p-k)=\frac{12i\pi^2g}{(g-2)(g^2-1)}\left(\frac{p}{\Lambda}\right)^{2\epsilon}, \quad (52)$$

$$R_3=\int\frac{d^4kp\cdot k[k\cdot(p-k)]^2}{k^8(p-k)^4}\left(\frac{k}{\Lambda}\right)^{2\epsilon}=\frac{i\pi^2}{2(g-1)(g-2)}\left(\frac{p}{\Lambda}\right)^{2\epsilon}, \quad (53)$$

where R_1, R_2, R_3 are evaluated in the manner used in the Appendix. Consequently,

$$f_3=\frac{-\alpha}{4\pi}\frac{(f_3-f)}{(g^2-1)}\left(2+4g+4g^2+\psi\frac{g^3}{g-2}\right) \quad (54)$$

or

$$\frac{f_3}{f}=\frac{\alpha[2+4g+4g^2+\psi g^3/(g-2)]}{4\pi(g^2-1)+\alpha[2+4g+4g^2+\psi g^3/(g-2)]}. \quad (55)$$

Having completed the calculation of f, g, ψ, and f_3, we substitute into Eq. (15) and evaluate F_1 as a function of α (or alternately g). The results will be discussed in Sec. IV.

IV. CONCLUSION

In this section we will show that our expression for F_1 reproduces low-order perturbation-theory results, discuss our solution in the light of gauge invariance, and discuss the singularity structure of F_1 as a function of α.

Our solutions will now be shown to agree with the results of low-order perturbation theory. We expand Eqs. (30) and (40) in a power series in the fine-structure constant and obtain

$$g=\frac{-\alpha}{4\pi}+\frac{1}{8}\left(\frac{\alpha}{2\pi}\right)^2+O(\alpha^3), \quad (56)$$

$$\psi=\frac{-\alpha}{4\pi}+O(\alpha^2), \quad (57)$$

if we choose the branch of the equations corresponding to the results of low-order perturbation theory. Substituting Eq. (55) into Eq. (15) yields

$$F_1 = \tfrac{2}{3}(1 - \tfrac{3}{2}g^2 - g^3) - \frac{\alpha(2+4g+4g^2)(g-2)+\alpha\psi g^3}{4\pi(g^2-1)(g-2)+\alpha(2+4g+4g^2)(g-2)+\alpha\psi g^3}. \tag{58}$$

Since Eqs. (56) and (57) imply $g=\psi=0$ at $\alpha=0$, we obtain

$$F_1 = \frac{2}{3} + \frac{\alpha}{2\pi} + (1+2g'-g'^2)\left(\frac{\alpha}{2\pi}\right)^2 + \cdots \tag{59}$$

when we expand around $\alpha = 0$, where

$$g' \equiv 2\pi \frac{dg}{d\alpha}\bigg|_{\alpha=0} = -\tfrac{1}{2}$$

by Eq. (56). Substituting for g' gives

$$F_1 = \frac{2}{3} + \frac{\alpha}{2\pi} - \frac{1}{4}\left(\frac{\alpha}{2\pi}\right)^2 + \cdots, \tag{60}$$

in agreement with known exact results of low-order perturbation theory. From Eq. (59) it is clear that terms in Eq. (60) are only sensitive to the $O(\alpha)$ behavior of g. This emphasizes the importance of the functional dependence of Eqs. (15) and (60) on g. This dependence results from three assumptions: (1) power-law behavior of Γ_μ, S_F, and $\Gamma_{\mu\alpha}$, (2) no contributions from terms involving the electron-positron kernel, K, and (3) the form of the vertex equation for $\Gamma_{\mu\alpha}$ given in Eq. (9).

The dependence of g upon α can be understood from the viewpoint of gauge invariance. From Eq. (13) we see

$$Z_2 \sim \Lambda^{2\epsilon}. \tag{61}$$

The generalized Landau gauge is defined[8] to be the gauge where Z_2 is finite. The photon propagator in this gauge differs from the Feynman gauge propagator by the term

$$\frac{G k_\mu k_\nu}{k^4}, \tag{62}$$

where $G = G(\alpha)$. Johnson and Zumino[9] have shown that a change of gauge by

$$D_{\alpha\beta} \to D_{\alpha\beta} - \gamma \frac{k_\alpha k_\beta}{k^2}\left(\frac{1}{k^2+\mu^2} - \frac{1}{k^2+\Lambda^2}\right) \tag{63}$$

changes Z_2 by

$$Z_2 \to Z_2 \left(\frac{\Lambda^2}{\mu^2}\right)^{(\alpha/4\pi)\gamma}. \tag{64}$$

In particular if we let $\gamma = -G + \psi$ and transform from the generalized Landau gauge to the gauge defined by ψ, we find

$$Z_2 \sim (\Lambda^2)^{(\alpha/4\pi)(\psi - G)} \tag{65}$$

in the ψ gauge. Using[8] $G = 1 - 3\alpha/8\pi$ and Eq. (57) for ψ gives

$$\frac{\alpha}{4\pi}(\psi - G) = \frac{\alpha}{4\pi}\left(\frac{-\alpha}{4\pi} - 1 + \frac{3\alpha}{8\pi}\right) \tag{66}$$

$$= \frac{-\alpha}{4\pi} + \frac{1}{8}\left(\frac{\alpha}{2\pi}\right)^2. \tag{67}$$

Comparing Eqs. (56) and (67) establishes

$$g = \frac{\alpha}{4\pi}(\psi - G) \tag{68}$$

in low order. Therefore our equations embody the general transformation properties of Z_2 under gauge transformation, at least in low order.

Having established the agreement of our work with low-order perturbation theory, we now discuss the behavior of our approximation for F_1. First, it should be noted that the solution for g given above (which agrees with low-order perturbation theory) is only one of several possible solutions. Figure 4 shows the relation between α and g and displays the different branches of $g(\alpha)$. The physical branch (i.e., the continuous part corresponding to perturbation-theory results) of $g(\alpha)$ has the range $[0.61, -0.34]$ with α varying from $-\infty$ to 3.4. The only singularity of F_1 on the physical branch is a zero at $\alpha \cong -3.7$. As Fig. 4 shows, the other branches have zeros at $\alpha \cong -2$ and -3.6, and singularities at $\alpha \cong 0.9$ and -3.06. F_1 is a rapidly varying function of α with a rich singularity structure in our approximation.

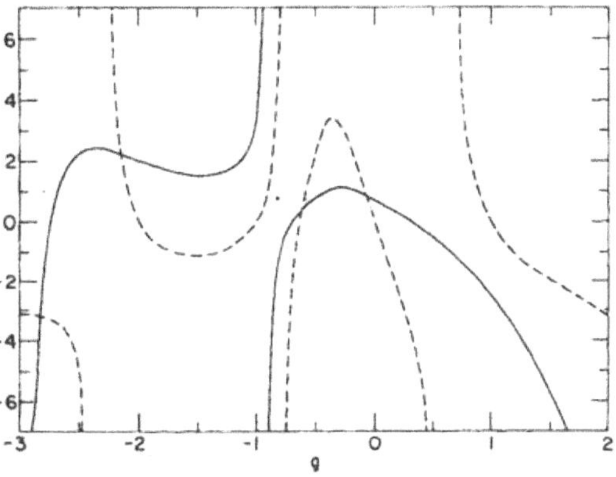

FIG. 4. The solid line is a parametric plot of F_1 versus g. The dashed line is a plot of α as a function of g. To find F_1 for a given α_0 draw a vertical line through the point α_0 on the $\alpha(g)$ curve. The intercept on the F_1 curve gives $F_1(\alpha_0)$ for the branch chosen.

At this point it should be remarked that other choices for the argument of the zero-momentum-transfer vertex functions in the kernel of Eq. (6), $\Gamma_\lambda\cdots\Gamma_\sigma$, such as $\Gamma_\lambda(p)\cdots\Gamma_\sigma(p)$ will not lead to expressions for F_1 with essential singularities and can lead to forms for F_1 in disagreement with the low-order results of Eq. (60).

In conclusion, we have summed the contributions of $(2N+1)\,(2^N-N)$ diagrams to the $O(\alpha^N)$ terms of F_1 and (after summation over N) found that a rapidly varying function of α with branch-point singularities results.

ACKNOWLEDGMENT

It is a pleasure to thank Professor Marshall Baker and other members of the University of Washington Physics Department for helpful discussions. In addition, I am grateful to the Aspen Center for Physics for its hospitality and to the members of the Field Theory Workshop for interesting conversations.

APPENDIX

In this appendix we outline the method we have followed in evaluating loop integrals in Sec. III. Our procedure is to make a Wick rotation and express the loop integral in spherical coordinates in a four-dimensional Euclidean space. The following identities are useful[10]:

$$\frac{1}{1+\alpha^2-2\alpha t}=\sum_0^\infty C_N^1(t)\alpha^N \tag{A1}$$

for $\alpha<1$ where C_N^1 is a Gegenbauer polynomial, and

$$\int d\Omega\, C_N^1 C_m^1 = 2\pi^2\delta_{Nm}. \tag{A2}$$

All nontrivial angular integrals encountered in Sec. III reduce to one of the following forms:

$$\int\frac{d\Omega}{(p-k)^2}=\frac{2\pi^2}{k_>^{\,2}}, \tag{A3}$$

$$\int\frac{d\Omega}{(p-k)^4}=\frac{2\pi^2}{k_>^{\,2}(k_>^{\,2}-k_<^{\,2})}, \tag{A4}$$

where $k_>$ ($k_<$) is the greater (lesser) of p and k.

We now will evaluate some integrals for the sake of illustration (we assume $g>0$ throughout):

$$I_1=\int\frac{d^4k}{(p-k)^2k^2}\left(\frac{k}{\Lambda}\right)^{2g} \tag{A5}$$

$$=2\pi^2 i\int_0^\Lambda\frac{dk\,k}{k_>^{\,2}}\left(\frac{k}{\Lambda}\right)^{2g} \tag{A6}$$

$$=2\pi^2 i\left[\int_0^p\frac{dk}{p^2}k\left(\frac{k}{\Lambda}\right)^{2g}+\int_p^\Lambda\frac{dk}{k}\left(\frac{k}{\Lambda}\right)^{2g}\right] \tag{A7}$$

$$=\frac{i\pi^2}{g}\left[1-\frac{1}{g+1}\left(\frac{p}{\Lambda}\right)^{2g}\right], \tag{A8}$$

which was given in Eq. (26). Our second example is

$$J_{0\mu\nu}=\int\frac{d^4k\,k_\mu k_\nu}{(p-k)^2k^4}\left(\frac{k}{\Lambda}\right)^{2g}. \tag{A9}$$

Now

$$J_{0\mu\nu}=J_{01}\,g_{\mu\nu}+J_{02}\frac{p_\mu p_\nu}{p^2} \tag{A10}$$

and as a result

$$J^{\mu}_{0\mu}=4J_{01}+J_{02}=I_1. \tag{A11}$$

Furthermore,

$$p_\mu p_\nu J_0^{\mu\nu}=p^2(J_{01}+J_{02}) \tag{A12}$$

$$=\int\frac{d^4k(k\cdot p)^2}{k^4(p-k)^2}\left(\frac{k}{\Lambda}\right)^{2g} \tag{A13}$$

$$=\frac{1}{4}\int\frac{d^4k}{k^4}\left(\frac{k}{\Lambda}\right)^{2g}\left[\frac{(p^2+k^2)^2}{(p-k)^2}-2(p^2+k^2)+(p-k)^2\right] \tag{A14}$$

$$=\frac{\pi^2 i}{2}\left[\int_0^p\frac{dk}{k}\left(\frac{k}{\Lambda}\right)^{2g}\left(k^2+\frac{k^4}{p^2}\right)+\int_p^\Lambda\frac{dk}{k}\left(\frac{k}{\Lambda}\right)^{2g}\left(p^2+\frac{p^4}{k^2}\right)\right] \tag{A15}$$

$$=\frac{\pi^2 i}{4}\left[\frac{p^2}{g}+\frac{2}{g(g+2)\,(g^2-1)}\left(\frac{p}{\Lambda}\right)^{2g}-\frac{4}{(g+2)\,(g-1)}\right] \tag{A16}$$

up to terms of $O(p^2/\Lambda^2)$, which we drop. Algebraic manipulation of Eqs. (A16) and (A10) gives

$$J_{01}=\frac{i\pi^2}{4}\left[\frac{1}{g}+\frac{2}{g(g+2)\,(g^2-1)}\left(\frac{p}{\Lambda}\right)^{2g}\right], \tag{A17}$$

$$J_{02}=\frac{-i\pi^2}{(g+2)\,(g-1)}\left(\frac{p}{\Lambda}\right)^{2g}. \tag{A18}$$

The above equation leads directly to Eq. (34).

*Work supported in part by the U.S. Atomic Energy Commission.

†Address after September 1, 1973: Physics Department, Cornell University, Ithaca, New York 14850.

[1]M. Baker and K. Johnson, Phys. Rev. D 3, 2516 (1971), and references contained therein; K. Johnson and M. Baker, *ibid.* 8, 1110 (1973).

[2]S. Adler, Phys. Rev. D 5, 3021 (1972).

[3]J. L. Rosner, Phys. Rev. Lett. 17, 1190 (1966); R. Jost and J. M. Luttinger, Helv. Phys. Acta 23, 201 (1950); A. E. Uehling, Phys. Rev. 48, 55 (1935).

[4]H. Brandt, thesis, University of Washington (Seattle) (unpublished). I am grateful to Professor M. Baker for drawing my attention to this article.

[5]K. Johnson, R. Willey, and M. Baker, Phys. Rev. 163, 1699 (1967).

[6]S. Blaha, Univ. of Washington report, 1972 (unpublished).

[7]M. Gell-Mann and F. Low, Phys. Rev. 95, 1300 (1954).

[8]K. Johnson, M. Baker, and R. Willey, Phys. Rev. 136, B1111 (1964).

[9]K. Johnson and B. Zumino, Phys. Rev. Lett. 3, 351 (1959).

[10]W. Magnus and F. Oberhettinger, *Formulas and Theorems for the Functions of Mathematical Physics* (Chelsea, New York, 1949).

PHYSICAL REVIEW D VOLUME 9, NUMBER 8 15 APRIL 1974

Higher-order calculation of transmission below the potential barrier

S. S. Wald and P. Lu

Department of Physics, Arizona State University, Tempe, Arizona 85281

(Received 26 December 1973)

In extending the Miller-Good modified WKB approximation to include the higher-order terms, a divergence was introduced. Because of this divergence, the approximation was limited to energies above the potential barrier. With this divergence removed, the modified WKB method is no longer limited to energies above the potential barrier. In order to demonstrate this method, we calculate the transmission coefficients for energies below the peak of the potential barrier and show that the higher-order terms are essential to the approximation.

I. INTRODUCTION

The conventional WKB approximation is widely known for its usefulness in solving simple barrier-penetration problems. However, as Ford *et al.*[1] pointed out, the conventional WKB method tends to break down as the energy approaches the potential-barrier top.

Miller and Good[2] proposed a modified WKB method in which the solutions of a model Schrödinger equation that can be solved exactly and resembles the actual Schrödinger equation would be used as the basis of the approximation. The reader is referred to their paper for details. However, the modified WKB method was only utilized to zeroth order in $\hbar^2$ because of divergences in the higher-order terms. Using the method developed by Lu and Measure[3] to remove the divergences in the higher-order terms, a divergence at the maximum point of the potential barrier was introduced. This divergence limits the approximation to energies above the barrier top where there are no real classical turning points, and hence there is no maximum point on the path of integration. Using the modified WKB method to first order in $\hbar^2$, we calculated the transmission coefficients above the potential barrier[4] and obtained agreement with the numerical results to at least four significant figures. This indicated how essential the higher-order terms are to the approximation.

For the case of penetration below the potential barrier, there are two classical turning points and one maximum point lying between the turning points. In order to remove the divergence at the maximum point, we start with the basic contour integral representation and then derive a formula which can be applied to the case of "penetration through the barrier." In Sec. II, this formula is derived in general terms. Using the Eckart potential as an example in Sec. III, we calculate the transmission coefficients for energies below the barrier, and the results are shown to be in agreement with the numerical results as presented in Table I. Thus the barrier-penetration problem can be solved using the modified WKB method with excellent results even for energies near the top of the potential barrier.

II. METHOD OF APPROXIMATION

In general, we wish to solve the Schrödinger equation

$$\left[\frac{d^2}{dx^2}+\frac{P_1^2(x)}{\hbar^2}\right]\psi(x)=0 \qquad (1)$$

Appendix B. "Asymptotic Form of the Electron Propagator and the Self-Mass of the Electron"

This paper is M. Baker and K. Johnson, Phys. Rev. **D3**, 2516 (1971). Reprinted with the kind permission of Physical Review D.

Asymptotic Form of the Electron Propagator and the Self-Mass of the Electron*

M. Baker
Physics Department, University of Washington, Seattle, Washington 98105

AND

K. Johnson
Laboratory for Nuclear Science and Physics Department, Massachusetts Institute of Technology, Cambridge, Massachusetts 02139

(Received 3 September 1970)

The electron self-mass problem is discussed in the context of ordinary renormalized quantum electrodynamics. All perturbation contributions to the renormalized self-energy part $\Sigma(p)$, which diverge logarithmically or remain constant in the limit when $p\gg m$, are summed. The resulting $\Sigma(p)$ vanishes in the limit $p^2/m^2\to\infty$ and yields a value for δm which is finite and equal to m. To obtain this result it is only assumed that the exact photon Green's function at small distances behaves like the bare propagator, which is the case if the eigenvalue equation for the bare coupling constant has a finite root. It is shown that in spite of the fact that the resulting mechanical mass m_0 vanishes identically, no conservation equation is obtained for any axial-vector current. Hence no Goldstone bosons appear in ordinary quantum electrodynamics when it is summed to all orders.

I. INTRODUCTION

In an earlier paper[1] we studied the unrenormalized Schwinger-Dyson equations for the electron propagator $S(p)$ under the assumption that the photon propagator $D(k)$ is proportional to $1/k^2$ as $k^2\to\infty$.[2] We found that if the electron bare mass m_0 was taken equal to be zero, we could obtain finite solutions for $S(p)$ in a certain approximation scheme. The resulting electron electromagnetic mass δm was finite and equal to the physical electron mass m, which of course was undetermined because the original $m_0=0$ equations contain no scale parameter.

The above work[1] was insufficient on two major counts:

(i) There was no explicit demonstration that these results would not be essentially modified as one went to higher orders in the approximation scheme.

(ii) The relation between the approximation scheme and the conventional renormalized perturbation expansion of quantum electrodynamics was not made clear.

The purpose of the present paper is to answer the questions raised by these points. We show [under the assumption that $D(k)\sim(1/k^2)$ as $k^2\to\infty$] that the ordinary renormalized perturbation solution for $S(p)$ sums to a function which, in an appropriately chosen gauge, has the following behavior as $p\to\infty$:

$$S^{-1}(p)\sim C^{-1}[\gamma\cdot p+am(m^2/p^2)^{\epsilon}]\,. \qquad (1.1)$$

ϵ is a constant which is determined by the expansion of the renormalized Bethe-Salpeter kernel K for electron-positron scattering. C and a are also constants. We explicitly calculate ϵ to order α_0^2 and find[3]

$$\epsilon=\frac{3}{2}\frac{\alpha_0}{2\pi}+\frac{3}{8}\left(\frac{\alpha_0}{2\pi}\right)^2+\cdots, \qquad (1.2)$$

where the unrenormalized fine-structure constant $\alpha_0\equiv e_0^2/4\pi$ is precisely defined in terms of the renormalized theory in Sec. II, Eq. (2.1).

The proof of (1.1) will make essential use of those properties of the scattering kernel K which were proven to all orders of perturbation theory in our discussion of vacuum polarization.[2] We obtain result (1.1) if the asymptotic behavior of the exact K is the same as the asymptotic behavior of the individual terms of its perturbation expansion which is the same in each order. Equation (1.1) then gives the exact asymptotic expression for the renormalized electron propagator $S(p)$ of quantum electrodynamics, provided that the photon propagator $D(k)\sim 1/k^2$ as $k^2\to\infty$.[2]

Furthermore, we show that the electromagnetic mass δm calculated in terms of the physical mass m is finite and equal to m for all values of the physical mass. The usual divergent expression for δm arises from using the perturbation-theory solution of $S(p)$ for high p^2 rather than the exact asymptotic solution given by (1.1).

In order to make the logic of the argument clear, we briefly summarize the basic outline of our approach, *ignoring* for clarity the technical difficulties associated

* This work is supported in part through the U. S. Atomic Energy Commission under Contract Nos. AT(30-1)-2098, AT(45-1)-1388B, and AT(30-1)-3829.

[1] K. Johnson, M. Baker, and R. Willey, Phys. Rev. **136** B1111 (1964).

[2] In two later papers [K. Johnson, R. Willey, and M. Baker, Phys. Rev. **163**, 1699 (1967); M. Baker and K. Johnson, *ibid.* **183**, 1292 (1969)] it was shown that a sufficient condition for the validity of the assumption $D(k)\to 1/k^2$ as $k^2\to\infty$ is the existence of a positive root of a certain equation $f(x)=0$. The function $f(x)$ is the coefficient of the logarithmic divergence in Z_3, calculated in the theory without photon self-energy insertions. See also M. Gell-Mann and F. E. Low, *ibid.* **95**, 1300 (1954).

[3] Equation (1.2) does not agree with the results obtained in Ref. 1. This is due to an incorrect treatment of electron self-energy insertions and Ward's identity in Ref. 1. See the end of Sec. IV of the present paper.

with gauge dependence and multiplicative renormalizations. Then, with these problems put aside, the function $S(p)$ is finite in perturbation theory, and

$$1/S(p)=\gamma\cdot p+m+\Sigma(p)-\delta m\,, \tag{1.3}$$

where m = the physical mass and the combination $\Sigma(p)-\delta m$ is finite. Formally $\Sigma(p)-\delta m=0$ when $\gamma\cdot p=-m$. However, this equation for δm will not be used. We first show that all the perturbation contributions to finite quantity $S^{-1}(p)-\gamma\cdot p$, which do not vanish in the limit when $p\gg m$, may be expressed in terms of $S^{-1}(p_0)-\gamma\cdot p_0$, where p_0 also is asymptotic. They will be related by an expression which does not involve the physical mass m. This is a nontrivial result and it is the reason that the analysis can be carried out. We next find that if we sum up all these nonvanishing perturbation contributions to $S^{-1}(p)-\gamma\cdot p$, the sum vanishes in the limit when $p/m\to\infty$. Thus, we shall obtain the nonperturbative result $S^{-1}(p)-\gamma\cdot p\to 0$ in the limit $p/m\to\infty$ or, because of (1.3),

$$m+\Sigma(p)-\delta m\to 0$$

in the limit when $p\gg m$. In this case the integrals which define the *unsubtracted* $\Sigma(p)$ when expressed in terms of the exact S converge and as a consequence we show that

$$\Sigma(p)\to 0\quad\text{for}\quad p\gg m\,. \tag{1.4}$$

When we combine (1.3) and (1.4), we find

$$m-\delta m\equiv 0\,. \tag{1.5}$$

Although the above paragraph basically describes our approach, the technical questions referred to make the actual calculations somewhat more involved. Hence one should not apply Eqs. (1.3) and (1.4) above without the qualifications which are appended to them in the sections which follow.

II. BEHAVIOR OF $S(p)$ FOR LARGE p^2

We want to study the high-p^2 behavior of $S(p)$ under the assumption that the renormalized photon propagator $D(k^2)$ behaves like $1/k^2$ as $k^2\to\infty$. We group together all those Feynman graphs for $S(p)$ which differ from each other only by insertions in internal photon lines. This grouping affects no conservation law of the exact theory; that is, the graphs in each group respect Lorentz invariance and current conservation. The sum of all graphs in each group is then equal to an equivalent single graph. In this equivalent graph, the internal photon-lines stand for $D(k^2)$ and the coupling constant on the end of each line is the renormalized charge e. If e_0^2, the unrenormalized charge, is finite, this equivalent graph has the same behavior in the high-p^2 region as the graph in which $D(k)$ is replaced by $1/k^2$ and e^2 by e_0^2 defined in the renormalized theory

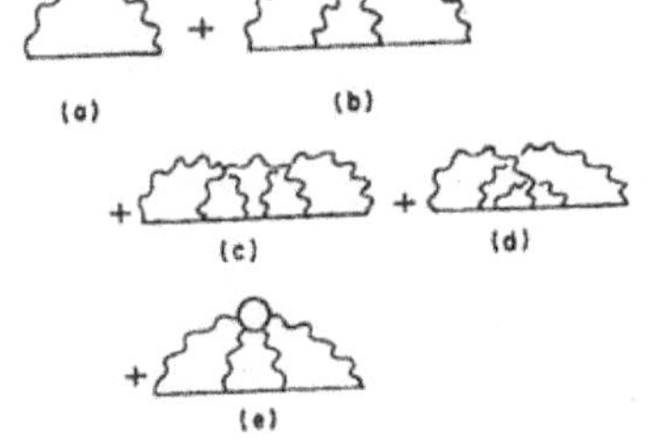

FIG. 1. Some typical diagrams for Σ^*.

by the equation

$$\lim_{k^2\to\infty} e^2k^2D(k)=e_0^2\,. \tag{2.1}$$

This is because the replacement of any single photon line in the graph by a contribution to $D(k)$ which vanishes more rapidly than $1/k^2$ makes all integrations converge. Because the graph is a function of only one external momentum p^2, this forces such a contribution to vanish as p^2 becomes large, as one can see from a simple scaling argument. Thus, if we assume that e_0^2 is finite,[4] we can calculate $S(p)$ in the uv region by omitting all graphs with photon self-energy insertions and by using the bare charge e_0 at the vertices.[5]

The sum of all such equivalent graphs yields an $S(p)$ which satisfies the functional equation

$$S^{-1}(p)=\gamma\cdot p+m_0+\Sigma^*(p\,;S(p'))\,, \tag{2.2}$$

where $m_0=m-\delta m$ is the bare mass of the electron. The functional $\Sigma^*(p\,;S(p'))$ is defined as the sum of all electron self-energy graphs which (a) cannot be broken by cutting a single electron line and (b) contain no insertions in either internal photon or electron lines. In each graph the internal electron lines stand for the full electron propagator $S(p)$, while the internal photon lines stand for the free photon propagator $D_{\mu\nu}{}^0(k)$, given by

$$D_{\mu\nu}{}^0(k)=\frac{1}{k^2}\left(g_{\mu\nu}+(b-1)\frac{k_\mu k_\nu}{k^2}\right), \tag{2.3}$$

where b is an arbitrary gauge parameter. Some of the typical low-order contributions to $\Sigma^*(p,S(p'))$ are depicted in Fig. 1.

The physical mass m of the electron is determined by

$$S^{-1}(\gamma\cdot p)=0\quad\text{for}\quad\gamma\cdot p=-m \tag{2.4a}$$

or, equivalently,

$$\delta m=\Sigma^*[p\,;S(p')]\quad\text{for}\quad\gamma\cdot p=-m\,. \tag{2.4b}$$

[4] The value of e_0^2 is the first positive root of the equation $f(e_0^2/8\pi^2)=0$, where $f(x)$ is defined in Ref. 2. To order x^2, $f(x)=\frac{3}{4}+x-\frac{1}{6}x^2$. See J. L. Rosner, Phys. Rev. Letters **17**, 1190 (1966).

[5] It has been shown (M. Baker and K. Johnson, Ref. 2) that if e_0^2 is finite then the leading correction to the asymptotic limit (2.1) for $D(k)$ is of the form $(1/k^2)(m^2/k^2)^{K(e_0^2)}$.

The iterative solution of (2.2) generates the usual unrenormalized perturbation expansion for $S(p)$. The divergences of the integrals which appear in this expansion can be isolated in terms of two infinite constants: the electron self-mass $\delta m=m-m_0$ and the electron-wave-function renormalization constant Z_2. However, it has been shown[6] that the perturbation expression for Z_2 is finite if the gauge constant b is suitably chosen. In this gauge the only infinities in the perturbation expansion of (2.2) arise from the δm divergences. Hence, if we subtract (2.2) at $\gamma\cdot p=-m$ and use (2.4b), we obtain

$$S^{-1}(p)=\gamma\cdot p+m+\Sigma^*(p;S(p')) -[\Sigma^*(p;S(p'))]_{\gamma\cdot p=-m}. \quad (2.5)$$

Because of the convenient choice of gauge, no second subtraction is necessary to render the perturbation solution of (2.5) finite. Thus from (2.5) we can obtain the usual renormalized perturbation expansion which expresses $S(p)$ in terms of m by a series of convergent integrals. The form for $S(p)$ in any other gauge can be obtained by a well-known transformation.[7]

The electromagnetic mass δm can be expressed in terms of $S(p)$ and hence m, using (2.4b). The resulting value of δm will be finite if $S(p)$ falls off sufficiently rapidly for $p^2\gg m^2$. Even if we ignore δm and the unrenormalized theory, and concern ourselves only with properties of the renormalized theory, the high-p^2 behavior is still of fundamental importance. For if $S^{-1}(p)$ contains terms which for large p^2 behave like $m[\ln(p^2/m^2)]$ as is indicated by renormalized perturbation theory, there then arises the possibility of inconsistencies in the renormalized theory when taken to all orders, i.e., the presence of "ghost" poles[8] in $S(p)$.

We will now show that the behavior of the solution of (2.5) at large p^2 is as indicated in (1.1). That is, we find

$$S(p)\to C(e_0^2)\left[\frac{1}{\gamma\cdot p}+\frac{m}{p^2}a_0(e_0^2)\left(\frac{m^2}{p^2}\right)^{\epsilon}\right], \quad (2.6)$$

where $C(e_0^2)$, $a_0(e_0^2)$, and $\epsilon(e_0^2)$ are constants to be defined below.

If we expand the $(m^2/p^2)^\epsilon$ factor in (2.6), we obtain the asymptotic expression for the usual renormalized perturbation-theory expansion, namely,

$$S(p)\to C(e_0^2)\left\{\frac{1}{\gamma\cdot p}+\frac{ma_0(e_0^2)}{p^2}\times\left[1-\epsilon\ln\frac{p^2}{m^2}+\tfrac{1}{2}\epsilon^2\left(\ln\frac{p^2}{m^2}\right)^2+\cdots\right]\right\}. \quad (2.7)$$

If expansion (2.7) for $S(p)$ is inserted in (2.4b), we obtain the usual perturbation series of logarithmically divergent integrals for δm. The logarithmic divergences arise from the terms proportional to ma_0/p^2 in (2.7). If instead we insert the correct exact asymptotic expression (2.6) for $S(p)$ in (2.4b), we obtain covergent integrals for δm because the factor $(m^2/p^2)^\epsilon$ with $\epsilon>0$ makes all logarithmically divergent integrals finite. Hence any discussion of self-mass integrals which involves the perturbation-theory estimate of the high-energy behavior of propagators is not relevant to the full theory.

Thus the basic problem is to show that the perturbation-theory logarithms sum to the form $(m^2/p^2)^\epsilon$ in (2.6). However, determination of the coefficient $C(e_0^2)$ of $1/\gamma\cdot p$ in (2.6) requires a bit of care and we will therefore make a few remarks about $C(e_0^2)$ before solving (2.5) for arbitrary m. If we set $m=0$, then the asymptotic solution (2.6) or (2.7) reduces to

$$S(p)^{m=0}=C(e_0^2)/\gamma\cdot p\,. \quad (2.8)$$

That is, we can calculate $C(e_0^2)$ by looking at ordinary perturbation theory with physical mass, $m=0$. The resultant integrals for $C(e_0^2)$ are finite because of our choice of gauge but, because of their superficial linear divergence, the value of the constant C depends upon the way the external momentum is routed through the diagrams and, further, the order of doing subintegrations.

One can determine C a little less ambiguously by calculating the electron-photon vertex function

$$\Gamma_\mu^{m\to0}(p,p+k)$$

and using Ward's identity,

$$\Gamma_\mu^{m=0}(p,p)=\frac{\partial}{\partial p^\mu}[S^{-1}(p)]^{m=0}=C^{-1}\gamma_\mu$$
$$=\gamma_\mu+\frac{\partial}{\partial p^\mu}\left[\Sigma^*\left(p;\frac{C}{\gamma\cdot p'}\right)\right]. \quad (2.9)$$

However, although the perturbation-theory integrals for Γ_μ may not be sensitive to the routing of the external momenta, they are not uniformly or absolutely covergent and still depend on the order of doing subintegrations. Different results for C can be obtained by introducing with different rules a cutoff Λ and then letting $\Lambda\to\infty$ for fixed p.[9] However, because the renormalized theory is free of ambiguities, the ambiguities associated with the different methods of defining

[6] K. Johnson, R. Willey, and M. Baker, Ref. 2.

[7] K. Johnson and B. Zumino, Phys. Rev. Letters 3, 351 (1959).

[8] These could arise in a manner similar to what might occur for the photon propagator if e_0^2 is infinite (see M. Baker and K. Johnson, Ref. 2).

[9] If we had introduced a cutoff Λ into the calculation of $S(p)$, then for p^2, $\Lambda^2\gg m^2$, we have $S(p)=(1/\gamma\cdot p)G(p^2/\Lambda^2,e_0^2)$. Now for fixed Λ^2, there are no uv divergences and the canonical commutation relations hold in their original form. This implies that as $p^2\to\infty$ for fixed Λ, $S(p)\to 1/\gamma\cdot p$, i.e., $G(\infty,e_0^2)=1$. However, since in our calculation no cutoff is introduced, we are effectively setting $\Lambda=\infty$ first. Hence we obtain $S(p)=(1/\gamma\cdot p)G(0,e_0^2)$, i.e., $C(e_0^2)=G(0,e_0^2)$. The fact that $G(0,e_0^2)$ is not necessarily equal to 1 and hence the fields have a modified canonical commutator reflects the sensitivity of the canonical commutation relations to the ambiguities of the perturbation theory integrals with no cutoff, even when there are no divergent quantities.

C cannot affect the value of any physical quantity. In particular the physically interesting gauge-invariant [see (5.17)] part of the asymptotic electron propagator, obtained by dividing out the factor C, will be independent of such ambiguities. For this reason, it is convenient to rewrite (2.2) in terms of a rescaled propagator $\tilde{S}(p)$ defined by the equation

$$\tilde{S}(p)=[1/C(e_0^2)]S(p)\,. \tag{2.10}$$

We could substitute (2.10) directly into the subtracted (2.5). However, since we are interested in the behavior of $S(p)$ for $p^2\gg m^2$, and since (2.5) involves the value of Σ^* at $\gamma\cdot p=-m$ explicitly via the subtraction term, it is instead more convenient first to subtract (2.2) at a value of $p=p^0$, where $(p^0)^2\gg m^2$. Then if we use the asymptotic solution of the resulting subtracted equation, we will be able to find the solution of (2.5) for $S(p)$ in terms of the physical mass m for $p^2\gg m^2$.

For this reason, instead of studying (2.5) directly, we will first rewrite (2.2) in terms of $\tilde{S}$ and then perform a subtraction at $p=p_0$. If we define $\tilde{m}(p)$ by the equation

$$\tilde{S}^{-1}(p)=\gamma\cdot p+\tilde{m}(p)\,, \tag{2.11}$$

(2.2) becomes

$$\tilde{m}(p)=Cm_0+(C-1)\gamma\cdot p+C\Sigma^*(p;C\tilde{S}(p'))\,. \tag{2.12}$$

We then subtract (2.12) at $p=p^0$ to obtain

$$\tilde{m}(p)=\tilde{m}(p_0)+(C-1)\gamma\cdot(p-p_0) +C[\Sigma^*(p;C\tilde{S}(p'))-\Sigma^*(p_0;C\tilde{S}(p'))]\,. \tag{2.13}$$

Equation (2.13) is an integral equation for $\tilde{m}(p)$. It contains as parameters the subtraction point p_0 and the subtraction constant $\tilde{m}(p_0)$. We first note that in the limit

$$\tilde{m}(p_0)\to 0\,, \tag{2.14}$$

(2.13) is satisfied if at the same time

$$\tilde{S}\to 1/\gamma\cdot p\,. \tag{2.15}$$

This is just a consequence of our definition (2.9) of C, as can be explicitly verified by inserting (2.14) and (2.15) into (2.13). This yields

$$0=(C-1)\gamma\cdot(p-p_0)+C\{\Sigma^*(p;C/\gamma\cdot p') -\Sigma^*(p;C/\gamma\cdot p')\}\,. \tag{2.16}$$

Differentiating (2.16) with respect to p_μ yields

$$0=\left(1-\frac{1}{C}\right)\gamma_\mu+\frac{\partial}{\partial p^\mu}\Sigma^*\left(p;\frac{C}{\gamma\cdot p'}\right),$$

which coincides with (2.9). Thus the p and p_0 terms in (2.16) vanish independently.

For the general case $\tilde{m}(p_0)\neq 0$, we make explicit the dependence of $\tilde{m}(p)$ upon p_0 and $\tilde{m}(p_0)$ by writing

$$\tilde{m}(p)=\tilde{m}(p_0)H(p,p_0;\tilde{m}(p_0))\,. \tag{2.17}$$

In Sec. III, we show that

$$\lim_{\tilde{m}(p_0)\to 0} H(p,p_0;\tilde{m}(p_0))=\text{finite}=H^a(p,p_0)\,, \tag{2.18}$$

i.e., the ratio $\tilde{m}(p)/\tilde{m}(p_0)$ approaches a finite limit as $\tilde{m}(p_0)$ approaches zero, for fixed p, p_0. It should be emphasized that (2.18) is a nontrivial statement and its truth is the basic reason that one can carry out this analysis.[10]

We will now show that result (2.18) allows us to calculate the exact asymptotic form for $\tilde{m}(p)$ for $p^2\gg m^2$. [$\tilde{m}(p)$ satisfies the rescaled version of the ordinary renormalized equation (2.5).] $\tilde{m}(p)$ has the form

$$\tilde{m}(p)=mF(p^2/m^2;e_0^2)+\gamma\cdot pG(p^2/m^2;e_0^2)\,, \tag{2.19}$$

where the functions F and G have the usual expansions in renormalized perturbation theory obtained by iterating (2.5) for S. For $p^2\gg m^2$, these expansions take the form

$$\begin{aligned}F(p^2/m^2)=&a_0(e_0^2)+a_1(e_0^2)\ln(p^2/m^2)\\&+a_2(e_0^2)[\ln(p^2/m^2)]^2+\cdots\\&+(m^2/p^2)[d_0+d_1\ln(p^2/m^2)+\cdots]\\&+\cdots\,,\end{aligned} \tag{2.20}$$

$$G(p^2/m^2)=(m^2/p^2)[b_0(e_0^2)+b_1(e_0^2)\times\ln(p^2/m^2)+\cdots]+\cdots\,. \tag{2.21}$$

Because of our choice of gauge and rescaling constant C, there are no terms in the asymptotic expansion (2.21) of G analogous to the $a_0+a_1\ln+a_2(\ln)^2+\cdots$ terms in expansion (2.20) of F.[11] Thus if we drop all those terms in the asymptotic expansion of the perturbation-series integrals for $\tilde{m}(p)$ which vanish as $p^2/m^2\to\infty$, we can write

$$\tilde{m}(p)\approx mF^a(p^2/m^2)\quad\text{for}\quad p^2\gg m^2\,, \tag{2.22}$$

where

$$F^a(p^2/m^2)=a_0(e_0^2)+a_1(e_0^2)\ln(p^2/m^2) +a_2(e_0^2)[\ln(p^2/m^2)]^2+\cdots\,. \tag{2.23}$$

Our problem is to sum the series (2.23), i.e., find a closed expression for the function $F^a(p^2/m^2)$. If we choose $p^2\gg m^2$ and $p_0^2\gg m^2$ and insert the asymptotic expression (2.22) for $\tilde{m}(p)$ and $\tilde{m}(p_0)$ into Eq. (2.17), we obtain the following functional equation for $F^a(p^2/m^2)$:

$$F^a(p^2/m^2)=F^a(p_0^2/m^2)H(p,p_0;mF^a(p_0^2/m^2))\,. \tag{2.24}$$

[10] The tautology (2.17) is of the same type that occurs in the so called "renormalization-group" analysis. However, that method has no content unless a zero-mass limit of the type (2.18) exists. This is made clear in the work of Gell-Mann and Low (Ref. 2), but it is not apparent in the work of many of the practioners of this method that an investigation of the sort carried out in (III) is required before one can believe its consequences.

[11] This is because one can verify that, to every order in perturbation theory, $m(p)=0$ when $m=0$. The argument is that used in the verification of (2.14) and (2.15). Thus G must vanish as $p^2/m^2\to\infty$, i.e., the asymptotic expansion of the perturbation-theory integrals for G must all have a factor m^2/p^2, as in (2.21).

Now let us assume that when $p^2/m^2 \to \infty$, F^a grows less rapidly than $(p^2/m^2)^{1/2}$, as indicated by perturbation theory; i.e., we assume

$$mF^a(p_0^2/m^2) \to 0 \tag{2.25}$$

as $m \to 0$. Then if we let $m \to 0$ in (2.24) and use the fundamental result (2.18), we obtain[12]

$$F^a(p^2/m^2) = F^a(p_0^2/m^2) H^a(p,p_0)\,, \tag{2.26}$$

which holds when $p^2/m^2 \gg 1$ and $p_0^2/m^2 \gg 1$.

Differentiating (2.26) with respect to p^2 and setting p_0^2 equal to p^2, we obtain

$$\frac{F^{a\prime}(p^2/m^2)}{F^a(p^2/m^2)} = -\epsilon \frac{m^2}{p^2}\,, \tag{2.27}$$

where

$$-\epsilon = p^2 \frac{d}{dp^2}[H^a(p,p_0)]_{p^2=p_0^2} \tag{2.28}$$

(H^a depends only on the ratio p^2/p_0^2). Hence

$$F^a(p^2/m^2) = A(m^2/p^2)^\epsilon\,. \tag{2.29}$$

Thus if $\epsilon > -\frac{1}{2}$, our assumption (2.25) is justified and (2.29) gives the exact asymptotic behavior of $\tilde{m}(p)$ for $p^2 \gg m^2$. In Sec. III, when we study the equation for $H^a(p,p_0)$, we will calculate ϵ to order e_0^4. The result is given by (1.2). The positivity of the first two terms in the power-series expansion then guarantees the validity of (2.29), at least for small e_0^2.

The constant A in (2.29) is clearly not determined from (2.26). However, (2.29) determines the values of all the constants $a_0, a_1, a_2, a_3, \ldots$ in expansion (2.23) in terms of any one of them [say, $a_0(e_0^2)$] and the constant ϵ; i.e., if we expand (2.29) in a power series in ϵ, we find

$$F^a(p^2/m^2) = A\{1 - \epsilon \ln(p^2/m^2) + \tfrac{1}{2}\epsilon^2[\ln(p^2/m^2)]^2 + \cdots\}\,. \tag{2.30}$$

Comparing (2.30) with (2.23), we obtain

$$A = a_0(e_0^2)\,. \tag{2.31}$$

Thus from (2.31), (2.29), and (2.22), we can write

$$\tilde{m}(p) \to m a_0(e_0^2)(m^2/p^2)^\epsilon\,, \quad p^2 \gg m^2\,. \tag{2.32}$$

Then using (2.10), (2.11), and (2.32), we obtain result (2.6) for the sum of all nonvanishing terms in the asymptotic expansion of the usual renormalized perturbation integrals for the electron propagator $S(p)$. (2.32) and (2.6) are valid in the gauge where Z_2 is finite, and include the contributions of all Feynman diagrams in which there are no photon self-energy insertions or, equivalently, of *all* diagrams if the photon self-energy insertions sum to the form $e^2 D_F(k) \to e_0^2/k^2$ with e_0^2 finite.

[12] Equation (2.26) implies that, for large p^2 and p_0^2, H^a depends only upon p^2 and p_0^2. This will be seen explicitly when we solve the equation for H^a.

It may seem that we have obtained the powerful result (2.32) without having made use of any properties of higher-order Feynman diagrams except for the general structure of their high-energy behavior [(2.20) and (2.21)]. However, the crucial ingredient for (2.32) was the assertion (2.18) that the ratio $\tilde{m}(p)/\tilde{m}(p_0)$ determined from (2.13) approached a finite limit as $\tilde{m}(p_0) \to 0$. The proof of this assertion requires that certain detailed and nontrivial properties of Feynman diagrams remain valid to every order in perturbation theory, as we shall see in Sec. III, where we derive (2.18).

III. DERIVATION OF EQ. (2.18)

When $\tilde{m}(p_0) = 0$, the solution of (2.13) is $\tilde{m}(p) = 0$. We can obtain an expression for $\tilde{m}(p)/\tilde{m}(p_0)$ by differentiating (2.13) with respect to $\tilde{m}(p_0)$ and setting $\tilde{m}(p_0) = 0$. This gives

$$\lim_{\tilde{m}(p_0)\to 0} \frac{\tilde{m}(p)}{\tilde{m}(p_0)} = \left.\frac{\partial \tilde{m}(p)}{\partial \tilde{m}(p_0)}\right|_{\tilde{m}(p_0)=0} = 1 + C\frac{\partial}{\partial \tilde{m}(p_0)}[\Sigma^*(p; C\tilde{S}(p')) - \Sigma^*(p_0, C\tilde{S}(p'))]_{\tilde{m}(p_0)=0}\,. \tag{3.1}$$

We now show that (3.1) yields a finite solution for

$$H^a = \lim_{\tilde{m}(p_0)\to 0} \frac{\tilde{m}(p)}{\tilde{m}(p_0)} \tag{3.2}$$

and hence we will establish our basic assertion (2.18). Now

$$\frac{\partial}{\partial \tilde{m}(p_0)}\{C\Sigma^*[p; C\tilde{S}(p'')]\}_{\tilde{m}(p_0)=0} = \int d^4p' C^2 \left\{\frac{\delta\Sigma^*[p; \tilde{S}(p'')]}{\delta\tilde{S}(p')}\right\}_{\tilde{m}(p_0)=0} \times \left[\frac{\partial\tilde{S}(p')}{\partial\tilde{m}(p_0)}\right]_{\tilde{m}(p_0)=0}\,. \tag{3.3}$$

But

$$-(2\pi)^4 \frac{\delta\Sigma^*(p, S(p''))}{\delta S(p')} \equiv K(p,p')\,, \tag{3.4}$$

where $K(p,p')$ is the Bethe-Salpeter kernel for electron position scattering.[13] We can understand (3.4) graphically

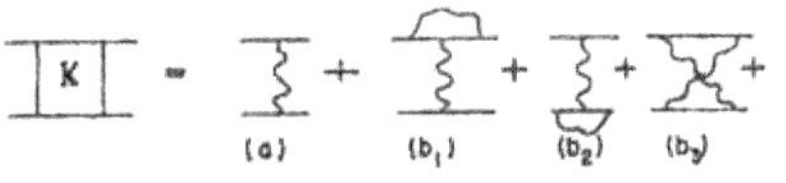

FIG. 2. Diagrams for K corresponding to diagrams (a) and (b) of Fig. 1.

[13] For a formal proof of (3.3), see for example, M. Baker, K. Johnson, and B. W. Lee, Phys. Rev. 133, B209 (1964).

ally by differentiating the contributions to Σ^* depicted in diagrams (a) and (b) of Fig. 1. The resulting diagrams for K are depicted in Fig. 2.

$K(p,p')$ can be expressed as a functional of S and the full vertex Γ_μ according to the expansion depicted in Fig. 3. From Fig. 3 it is clear that $K(S,\Gamma)$ satisfies the following scaling property:

$$\tilde{K}=C^2K(S,\Gamma)=K(\tilde{S},\tilde{\Gamma})\,, \qquad (3.5)$$

where

$$\tilde{\Gamma}_\mu=C\Gamma_\mu\,. \qquad (3.6)$$

Thus $\tilde{K}\equiv C^2K$ can also be represented by the expansion depicted in Fig. 3, provided we interpret that intermediate electron lines and vertex blobs in that diagram as representing $\tilde{S}$ and $\tilde{\Gamma}$, respectively.

In (3.3) we need $\tilde{K}$ evaluated for $\tilde{m}(p_0)=0$. Let us call

$$\tilde{K}|_{\tilde{m}(p_0)=0}\equiv\tilde{K}^a(p,p')\,.$$

From (3.5), (2.14), and (2.15), we see that

$$\tilde{K}^a(p,p')=K(p,p';1/\gamma\cdot p'',\tilde{\Gamma}^a)\,, \qquad (3.7)$$

where $\tilde{\Gamma}_\mu{}^a$ is the value of $\tilde{\Gamma}_\mu$ in (3.6) at $\tilde{m}(p_0)=0$.

If any of the integrals in the perturbation expansion for $\tilde{K}^a$ diverged, then (3.1) would make no sense and $\lim\tilde{m}(p)/\tilde{m}(p_0)$ as $\tilde{m}(p_0)\to 0$ would not exist. However, we have already investigated the properties of $\tilde{K}^a$ in our previous discussion of the Bethe-Salpeter equation for the vertex function $\tilde{\Gamma}$,[14]

$$\tilde{\Gamma}_\mu=C\gamma_\mu+\tilde{K}\tilde{S}\tilde{\Gamma}_\mu\tilde{S}\,. \qquad (3.8)$$

We showed that $\tilde{K}^a(p,p')$ is finite to all orders in perturbation theory[15] and *furthermore* no infrared divergences arise when we set $p=0$; i.e., $\tilde{K}^a(0,p')$ is also finite. The latter property allowed us to choose the gauge constant b so that (3.8) has a finite iteration solution or, equivalently, so that Z_2 is finite to all orders in perturbation theory.[16] We now see that the finiteness of $\tilde{K}^a(0,p)$, which was essential to our previous discussion of Z_2 and Z_3, also guarantees the finiteness of the solution H^a of (3.1). We can put this

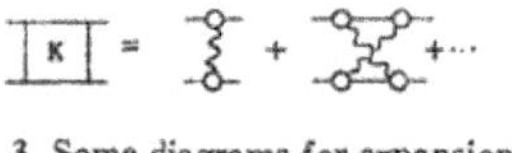

FIG. 3. Some diagrams for expansion of K in terms of the full vertex Γ.

[14] K. Johnson, R. Willey, and M. Baker, Ref. 2, Sec. IV and Appendix.

[15] Analogous properties of $\tilde{K}^a$ play an essential role in our discussion of Z_3. See K. Johnson, M. Baker, and R. Willey and M. Baker and K. Johnson (Ref. 2).

[16] The gauge constant b is determined by the following condition:

$$\int d\Omega'\tilde{K}^a(0,p')\frac{1}{\gamma\cdot p'}\gamma_\mu\frac{1}{\gamma\cdot p'}=0.$$

See Ref. 14.

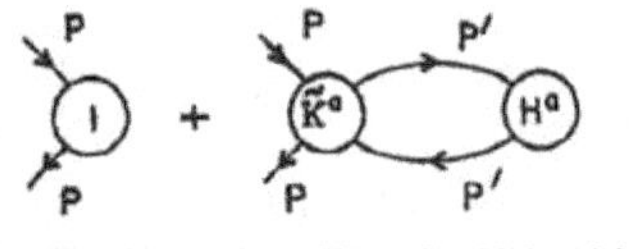

FIG. 4. Graphical representation of (3.10). The blob with K, stands for difference $\tilde{K}^a(p,p')-\tilde{K}^a(p_0,p')$.

equation in its final form by noting

$$\frac{\partial\tilde{S}(p')}{\partial\tilde{m}(p_0)}\bigg|_{\tilde{m}(p_0)=0}=-\tilde{S}(p')\frac{\partial\tilde{m}(p')}{\partial\tilde{m}(p_0)}\tilde{S}(p')\bigg|_{\tilde{m}(p_0)=0}$$
$$=-\frac{1}{\gamma\cdot p'}\frac{\partial\tilde{m}(p')}{\partial\tilde{m}(p_0)}\bigg|_{\tilde{m}(p_0)=0}\frac{1}{\gamma\cdot p'}. \qquad (3.9)$$

Combining (3.1)–(3.3) and (3.9), we obtain

$$\frac{\tilde{m}(p)}{\tilde{m}(p_0)}\bigg|_{\tilde{m}(p_0)\to 0}=H^a(p,p_0)$$
$$=1+\int\frac{d^4p'}{(2\pi)^4}[\tilde{K}^a(p,p')-\tilde{K}^a(p_0,p')]$$
$$\times\frac{1}{\gamma\cdot p'}H^a(p',p_0)\frac{1}{\gamma\cdot p'}. \qquad (3.10)$$

Equation (3.10) is depicted graphically in Fig. 4. The kernel $\tilde{K}^a$ is defined by integrals with zero-mass internal electron lines, (3.7), and those zero-mass integrals might have diverged in the infrared region. The demonstration that such infrared divergences do not arise in any order of perturbation theory was the essential part of our previous proof[14] that $\tilde{K}^a(p,p')$ is finite.

We now see how our previous analysis of zero-mass Feynman integrals not only guarantees that the kernel in (3.10) exists, but also guarantees the existence of a solution of (3.10) to all orders in perturbation theory. The first iteration of (3.10) yields the integral[17]

$$I(p,p_0)=-\int\frac{d^4p'}{(2\pi)^4}[\tilde{K}^a(p,p')-\tilde{K}^a(p_0,p')]\frac{1}{(p')^2}. \qquad (3.11)$$

We can choose the vectors p and p_0 in (3.10) and (3.11) to be spacelike so that we can rotate the contour in the p'^0 integration. We then write

$$\int\frac{d^4p'}{(2\pi)^4}=\frac{i}{16\pi^2}\int(p')^2d(p')^2\int\frac{d\Omega'}{2\pi^2}, \qquad (3.12)$$

[17] We have suppressed the indices of the Dirac matrices appearing in (3.10) and (3.11); e.g., the 1 in (3.10) stands for the Dirac matrix unity. With indices included, (3.10) becomes

$$I_{\alpha\beta}(p,p_0)=-\sum_\gamma\int\frac{d^4p'}{(2\pi)^4}[\tilde{K}_{\alpha\beta,\gamma\gamma}{}^a(p,p')-\tilde{K}_{\alpha\beta,\gamma\gamma}{}^a(p_0,p')]\frac{1}{p'^2}.$$

Since $\tilde{K}^a$ contains an even number of γ matrices, $I_{\alpha\beta}$ cannot contain a term like $(\gamma\cdot p)_{\alpha\beta}$ or $(\gamma\cdot p_0)_{\alpha\beta}$ and hence must be proportional to the unit Dirac matrix.

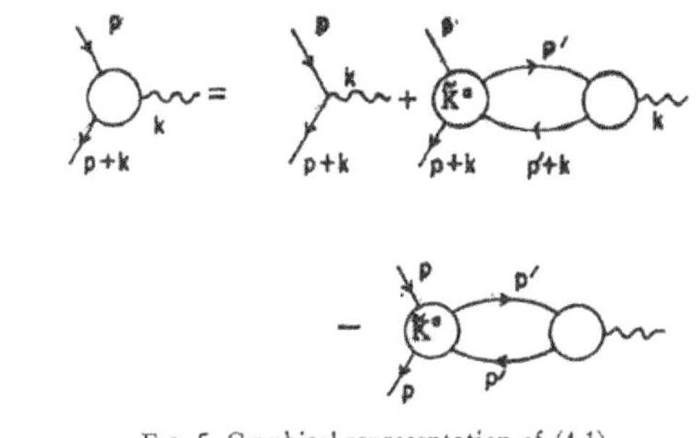

FIG. 5. Graphical representation of (4.1). Each vertex stands for $\tilde{\Gamma}^a$.

where the solid-angle integration in (3.12) is carried out over a sphere in four-dimensional Euclidean space. Since $\tilde{K}^a$ has dimensions $1/p^2$, we can write

$$\frac{i}{16\pi^2}\int\frac{d\Omega}{2\pi^2}\tilde{K}^a=\frac{1}{p'^2}k\left(\frac{p^2}{p'^2}\right), \tag{3.13}$$

where $k(p^2/p'^2)$ is a dimensionless function of p^2/p'^2. Equation (3.11) can then be written

$$I=-\int_0^\infty\frac{dp'^2}{p'^2}\left[k\left(\frac{p^2}{p'^2}\right)-k\left(\frac{p_0^2}{p'^2}\right)\right]. \tag{3.14}$$

Now if $k(p^2/p'^2)$ contained terms which behaved like $\ln(p'^2/p^2)$ as $p'^2\to\infty$, the subtraction in Eq. (3.14) would not make the integral I converge in the high-p'^2 region. However, from definition (3.13) and our result that $\tilde{K}^a(0,p)$ is finite, it follows that

$$\lim_{p'^2\to\infty}\frac{1}{p'^2}k\left(\frac{p^2}{p'^2}\right)\to\frac{k(0)}{p'^2}, \tag{3.15}$$

where $k(0)$ is finite constant. Thus the subtraction in (3.14) produces convergence in the high-p'^2 region.

From (3.15) and the symmetry property

$$\frac{1}{p'^2}k\left(\frac{p^2}{p'^2}\right)=\frac{1}{p^2}k\left(\frac{p'^2}{p^2}\right), \tag{3.16}$$

it also follows that the integrals in (3.14) converge in the low-p'^2 region even though we have set $m=0$.[18]

Thus we conclude that the first iteration of (3.10) yields a convergent integral I, which is a function $I(p_0^2/p^2)$ of the ratio p_0^2/p^2. Since $\tilde{K}^a(0,p')$ is finite,[19] $I(p_0^2/p^2)$ diverges logarithmically at $p^2\to 0$ and thus behaves (to within logarithms) like the inhomogeneous term 1 as $p^2\to 0$. Hence the higher-order iterations of (3.10) will also yield convergent integrals for $H^a(p,p_0)=H^a(p^2/p_0^2)$.[20]

We have therefore shown that our previous analysis[14] of the zero-mass integrals for K^a is sufficient to guarantee existence of the function $H^a(p^2/p_0^2)$ to every order in perturbation theory. Thus, without any new analysis of Feynman graphs, we have derived the basic equation (2.18) from which the asymptotic behavior of $\tilde{m}(p)$, (2.32), follows. The constant ϵ can be calculated from (2.28) by using the solution of (3.10) for $H(p^2/p_0^2)$ as a power series in e_0^2. However, it is more convenient and illuminating to calculate ϵ directly from the exact solution of (3.10) instead of using the iterative solution. With (3.12) and (3.13) and the fact that H^a depends only upon the ratio of p^2 and p_0^2, we can write (3.10) in the form

$$H^a\left(\frac{p^2}{p_0^2}\right)=1-\int_0^\infty\frac{dp'^2}{p'^2}H^a\left(\frac{p'^2}{p_0^2}\right)\times\left[k\left(\frac{p^2}{p'^2}\right)-k\left(\frac{p_0^2}{p'^2}\right)\right]. \tag{3.17}$$

From (2.26) and (2.29), we see that a solution has the form

$$H^a(p^2/p_0^2)=(p_0^2/p^2)^\epsilon. \tag{3.18}$$

The form (3.18) of the solution can also be directly obtained from (3.17). The integrals on the right-hand side of (3.17) will converge near $p'^2=0$ provided $\epsilon<1$. There will be convergence in the high-p^2 region provided $\epsilon>-1$. However, (2.26), from which we determined $\tilde{m}(p)$ [(2.32)], will be valid only if ϵ turns out to be $>-\frac{1}{2}$. If we substitute (3.18) into (3.17) and use the symmetry property (3.16), we obtain the following equation determining ϵ in terms of k:

$$-1=\frac{k(0)}{\epsilon}+\int_0^1 du\,k(u)\frac{1}{u^\epsilon}+\int_0^1 du\frac{k(u)-k(0)}{u^{1-\epsilon}}, \tag{3.19}$$

which is valid for $-1<\epsilon<1$. If $\epsilon>0$ then (3.19) may be put in the form

$$-1=\int_0^1 du\,k(u)(u^{-\epsilon}+u^{\epsilon-1}), \tag{3.20}$$

and in this case H^a then satisfies the *homogeneous* equation

$$H^a\left(\frac{p^2}{p_0^2}\right)=-\int_0^\infty\frac{dp'^2}{p'^2}H^a\left(\frac{p'^2}{p_0^2}\right)k\left(\frac{p^2}{p'^2}\right). \tag{3.21}$$

The $u^{-\epsilon}$ term in (3.20) arises from the p'^2 integration in the region $p'^2<p^2$ in (3.21) while the $u^{\epsilon-1}$ term is the contribution of the region $p'^2>p^2$.

[18] Equation (3.14) would of course converge at low p'^2 even if $k(p'^2/p^2)\sim\ln(p^2/p'^2)$, as $p'^2\to 0$. That is, the absence of such logarithms is only necessary for the high-p'^2 behavior of the integral.

[19] Setting $p=0$ in the integrals for $K^a(p,p')$ enhances the possibility of infrared divergences. In fact, our previous analysis of these integrals showed that the general diagram had just enough factors to guarantee convergence in the infrared region.

[20] We will give the explicit form of the higher-order interactions below when we calculate the exact solution of (3.10).

Thus we have shown that (3.10) has the finite solution (3.18), where ϵ is determined by (3.19) or (3.20). If the resulting value of ϵ lies between $-\frac{1}{2}$ and 1, then the assumptions (2.18) and (2.25) are justified and the proof of our basic result (2.32) is completed. In Sec. IV we calculate the first two terms in the power-series expansion of $\tilde{K}^a$ so that we can determine ϵ to order e_0^4 from (3.20).

IV. CALCULATION OF ϵ TO ORDER e_0^4

We first calculate k using its definition (3.13) in terms of $\tilde{K}^a$. From (3.7), $\tilde{K}^a(p,p')=K(1/\gamma\cdot p,\tilde{\Gamma}_\mu{}^a)$, where the functional K is represented by the series of diagrams depicted in Fig. 3. Thus to calculate $\tilde{K}^a$, we simply replace each internal photon line by $D_{\mu\nu}{}^0(k)$ [(2.3)], each internal electron line by $1/\gamma\cdot p$, and each vertex blob $\tilde{\Gamma}_\mu{}^a$. $\tilde{\Gamma}_\mu{}^a$ is determined by (3.8) with m set equal to zero. Since by choice of C, $\tilde{\Gamma}_\mu{}^a(pp)=\gamma_\mu$, we can determine C by setting the photon momentum $k=0$ in (3.8). If we insert the resulting expression for C in (3.8), we obtain the following equation for $\tilde{\Gamma}_\mu{}^a(p,\,p+k)$:

$$\tilde{\Gamma}_\mu{}^a(p,\,p+k)=\gamma_\mu+\int\frac{d^4p'}{(2\pi)^4}\tilde{K}^a(p,\,p+k;\,p',\,p'+k)$$

$$\times\frac{1}{\gamma\cdot p'}\tilde{\Gamma}_\mu(p',\,p'+k)\frac{1}{\gamma\cdot(p'+k)}$$

$$-\int\frac{d^4p'}{(2\pi)^4}\tilde{K}^a(p,p;\,p',p')\frac{1}{\gamma\cdot p'}\tilde{\Gamma}_\mu(p',p')\frac{1}{\gamma\cdot p'}. \quad (4.1)$$

Equation (4.1) is depicted graphically in Fig. 5. To lowest order,

$$\tilde{K}^{a(2)}=-ie_0{}^2\gamma^a\left(g^{ab}-\frac{(p-p')^a(p-p')^b}{(p-p')^2}\right)\gamma^b\frac{1}{(p-p')^2}; \quad (4.2)$$

the gauge constant b in (4.2) was chosen so that the condition in Ref. 16 is satisfied. To this order this gives $b=0$; thus, as is well known that the second-order vertex is finite in the Landau gauge. From (4.1) and (4.2), we then obtain the e_0^2 contribution to $\tilde{\Gamma}_\mu{}^a(p,\,p+k)$,

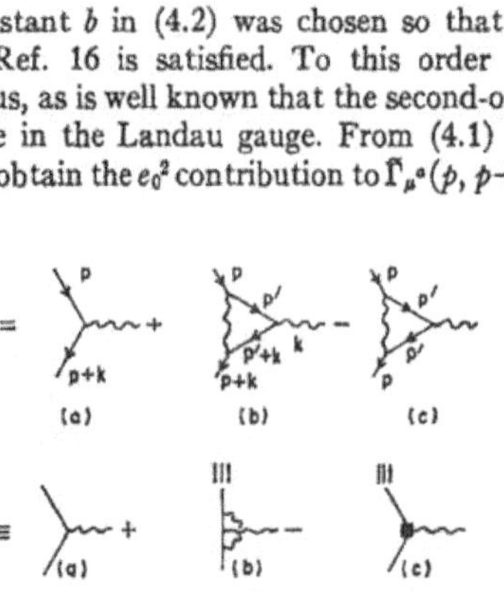

FIG. 6. Graphs for $\tilde{\Gamma}_\mu{}^a$ to order e_0^2. The second representation is just a convenient abbreviated notation.

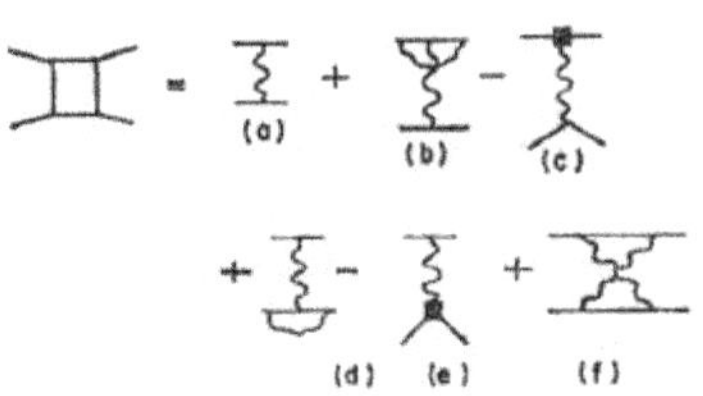

FIG. 7. Graphs for $\tilde{K}^a$ to order e_0^4.

which is depicted graphically in Fig. 6. The e_0^2 contribution to $\tilde{K}^a$ is then obtained by inserting the diagrams of Fig. 6 in the vertex blobs of Fig. 3. The resulting diagrams for the e_0^4 contribution to $\tilde{K}^a$ are shown in Fig. 7. Graph (a) of Fig. (7) includes a contribution of order e_0^4 arising from the e_0^2 term in the gauge constant b given by[1]

$$B^{(2)}=3\alpha_0/8\pi\,. \quad (4.3)$$

To order e_0^2 we find, using (4.2) and (3.12),

$$k(u)=-3\alpha_0/4\pi\,,\quad u\leqslant 1\,. \quad (4.4)$$

If we insert (4.4) in (3.19) or (3.20), we find

$$1=\frac{3\alpha_0}{4\pi}\left(\frac{1}{1-\epsilon}+\frac{1}{\epsilon}\right)+\cdots, \quad (4.5)$$

where the $1/\epsilon$ term arises from the second term in (3.20), which in turn comes from the large-P'^2 integration in (3.17). From (4.5), we find

$$\epsilon=3\alpha_0/4\pi+\cdots\,. \quad (4.6)$$

The higher-order terms in the expansion of ϵ in a power series in α_0 arise both from the contribution of the $1/(1-\epsilon)$ term in (4.5) and from the higher-order corrections to $\tilde{K}^a$ depicted in Fig. 7. However, in order to calculate ϵ to order α_0^2 we need only evaluate the α_0^2 contribution to $k(u)$ at $u=0$. This is easily seen if we write (3.19) in the form

$$-\epsilon=k(0)+\epsilon\int_0^1 du\,k(u)u^{-\epsilon}$$

$$+\epsilon\int_0^1 du\,u^{\epsilon-1}[k(u)-k(0)]. \quad (4.7)$$

Since the integrals on the right-hand side of (4.7) are finite as $\epsilon\to 0$, they give a contribution which is of $\alpha_0\times$(first-order term). To order α_0^2, (4.7) becomes

$$\epsilon^{(4)}=-k^{(4)}(0)+(3\alpha_0/4\pi)^2\,. \quad (4.8)$$

We can calculate $k^{(4)}(0)$ by setting $p=0$ in the integrals represented by Fig. 7. The calculation is straightforward and the contributions of the various diagrams of Fig. 7 to $k^{(4)}(0)$ are displayed in Table I. Using (4.3), we sum all the contributions listed in Table I. This

TABLE I. Contributions of diagrams of Fig. 7 to $k^{(4)}(0)$.

(a)	(b)	(c)	(d)	(e)	(f)
$-3(\alpha_0/4\pi)-b^{(2)}(\alpha_0/4\pi)$	$\frac{3}{2}(\alpha_0/4\pi)^2$	$\frac{9}{2}(\alpha_0/4\pi)^2$	$\frac{3}{2}(\alpha_0/4\pi)$	$\frac{9}{2}(\alpha_0/4\pi)^2$	$-3(\alpha_0/4\pi)^2$

yields

$$k^{(4)}(0)=-\frac{3\alpha_0}{4\pi}+\frac{15}{2}\left(\frac{\alpha_0}{4\pi}\right)^2. \qquad (4.9)$$

Equations (4.4) and (4.9) then give

$$\epsilon^{(4)}=\frac{3\alpha_0}{4\pi}+\frac{3}{2}\left(\frac{\alpha_0}{4\pi}\right)^2 \qquad (4.10)$$

as stated in Sec. I. Thus we see that at least for small α_0, ϵ is positive, and $S(p)$ behaves as indicated in (1.1) for large p^2. In Ref. 1, we overlooked the contribution of diagrams (c) and (e) of Fig. 7 to $\tilde{K}^a$. These diagrams arise from the vertex subtraction in (4.1) for $\bar{\Gamma}_\mu{}^a$ [diagram (c) of Fig. 6]. The omission was due to an incorrect treatment of Ward's identity given there and thus our previous result[1] for ϵ differed from (4.10) in the order α_0^2.

V. SHORT-DISTANCE BEHAVIOR OF ELECTRON PROPAGATOR IN ARBITRARY GAUGE

Our result (2.6) for the large-p behavior of $S(p)$ is valid in the gauge in which Z_2 is finite. To order α_0 this gauge is determined by (4.3). In any other gauge, the unrenormalized $S(p)$ is infinite and hence one must introduce a cutoff Λ in order to define it. We can do this by replacing the photon propagator $D_{\mu\nu}{}^0(k)$ by $D_{\mu\nu}{}^0(k,\Lambda)$, where

$$D_{\mu\nu}{}^0(k,\Lambda)=\frac{\Lambda^2}{\Lambda^2+k^2}D_{\mu\nu}{}^0(k). \qquad (5.1)$$

This makes the electron propagator $S_b(p,\Lambda)$, calculated using the photon propagator (5.1) for internal photon lines, finite in any gauge. We have introduced the subscript b to make explicit the dependence upon the gauge parameter. Then one can relate the *coordinate space* electron propagagor in the gauges b_1 and b_2 according to the formula[7]

$$S_{b_2}(x-x',\Lambda)=S_{b_1}(x-x',\Lambda)\times\exp\{ie_0^2(b_2-b_1)[D^0(x-x',\Lambda)-D^0(0,\Lambda)]\}, \qquad (5.2)$$

where

$$D^0(x,\Lambda)=-\int\frac{d^4k}{(2\pi)^4}\frac{e^{ik\cdot x}}{(k^2-i\epsilon)^2}\frac{\Lambda^2}{\Lambda^2+k^2} \qquad (5.3)$$

and

$$S_b(x-x',\Lambda)\equiv\int\frac{d^4p}{(2\pi)^4}e^{ip\cdot(x-x')}S_b(p,\Lambda). \qquad (5.4)$$

We will choose b_1 so that $S_{b_1}(x-x',\Lambda)$ is finite as $\Lambda\to\infty$, i.e.,

$$\lim_{\Lambda\to\infty}S_{b_1}(p,\Lambda)=S(p), \qquad (5.5)$$

where for large p, $S(p)$ is given by (2.6). This means that

$$b_1=3\alpha_0/8\pi+\cdots. \qquad (5.6)$$

We let b_2 be arbitrary and define

$$\delta=b_2-b_1. \qquad (5.7)$$

Then in the limit of large Λ, (5.2) becomes

$$S_{b_2}(x-x',\Lambda)=S(x-x')\exp[ie_0^2\delta I(x-x',\Lambda)], \qquad (5.8)$$

where

$$I(x-x',\Lambda)=\int\frac{d^4k}{(2\pi)^4}\frac{e^{ik\cdot(x-x')}-1}{(k^2-i\epsilon)^2}\left(\frac{\Lambda^2}{\Lambda^2+k^2}\right) \qquad (5.9)$$

and $S(x-x')$ is the Fourier transform (5.4) of $S(p)$ [(5.5)]. It is clear since $S(x-x')$ remains finite as $\Lambda\to\infty$, that in order to define a finite S_{b_2} an infinite renormalization is required. The resulting finite renormalized S_{b_2} will be related to $S(x-x')$ by a *factor* which is a finite function of $x-x'$, and which is so smooth that no divergences in the Fourier transform are produced by the confluence of the light cone singularities of $S(x-x')$ and the factor. We notice that since this factor is independent of spin that the coefficients of the Dirac matrix $\gamma\cdot(x-x')$ and 1 in S is the same, that is the ratio of these coefficients is gauge invariant, and is furthermore independent of the way the multiplicative renormalization is carried out. Therefore, we write

$$S(x-x')=\frac{\gamma\cdot(x-x')}{(x-x')^4}A(x-x')+m\frac{B(x-x')}{(x-x')^2}$$

and in an arbitrary gauge b,

$$S_b{}^{\text{ren}}(x-x')=\frac{\gamma\cdot(x-x')}{(x-x')^4}A_b(x-x')+m\frac{B_b(x-x')}{(x-x')^2}.$$

We can calculate S at short distances by using the large-p behavior of $S(p)$, i.e.,

$$S(x-x')\big|_{(x-x')^2\to0}\to C(e_0^2)\int\frac{dp}{(2\pi)^4}e^{ip\cdot(x-x')}\times\left[\frac{1}{\gamma\cdot p}+\frac{m}{p^2}a_0\left(\frac{m^2}{p^2}\right)^\epsilon\right]=\frac{C(e_0^2)}{2\pi^2}\left\{\frac{\gamma\cdot(x-x')}{(x-x')^4}+i\frac{\Gamma(1-\epsilon)}{2^{1+2\epsilon}\Gamma(1+\epsilon)}a_0\frac{m}{(x-x')^2}[m^2(x-x')^2]^\epsilon\right\};$$

the ratio

$$R(x-x')=\frac{mB(x-x')}{A(x-x')}=m\frac{B_b(x-x')}{A_b(x-x')}$$

is gauge invariant and independent of the method of carrying out the multiplicative renormalization of S_b. At small distances, R takes the form

$$R\to mi\frac{\Gamma(1-\epsilon)}{2^{1+2\epsilon}\Gamma(1+\epsilon)}a_0(e_0^2)[m^2(x-x')^2]^\epsilon.$$

R is a gauge-invariant finite quantity which at small distances contains functions of the charge: $a_0(e_0^2)$ and $\epsilon(e_0^2)$, which are expressed in terms of unambiguous integrals and which occur in renormalized perturbation theory. Note that the constant $C(e_0^2)$, which contains ambiguities, does not occur in $R(x-x')$.

VI. CALCULATION OF δm

We now insert our result for $S(p')$, expressed in terms of the physical mass m, in the original unrenormalized (2.2). In the appropriate gauge (5.6), all the integrals defining $\Sigma^*(p,S(p'))$ now converge since $S(p')$ possesses the large-p behavior given by (2.6). Hence $\Sigma^*(p,S(p'))$ can be expressed in terms of finite functions of the ratio (p^2/m^2). We can then determine the high-p behavior of Σ^* by letting m approach zero. The high-p limit of (2.2) yields the following equation for δm, or $m_0=m-\delta m$:

$$C^{-1}\gamma\cdot p=\gamma\cdot p+m_0+\Sigma^*(p,C/\gamma\cdot p')$$
$$-m\int\frac{d^4p'}{(2\pi)^4}K(p,p')|_{m\sim 0}\frac{\partial S(p')}{\partial m}\bigg|_{m\sim 0} \tag{6.1}$$

as $p/m\to\infty$. In arriving at (6.1) we have used the definition (3.4) of K and the fact that $S(p')|_{m=0}=C/\gamma\cdot p'$. By Ward's identity (2.9) the γp terms in (6.1) are canceled by the $\Sigma^*(p,\,C/\gamma\cdot p')$ term. Equation (6.1) then becomes

$$\delta m=m-m\int\frac{d^4p'}{(2\pi)^4}K(p,p')|_{m=0}\frac{\partial S(p')}{\partial m}\bigg|_{m\sim 0}. \tag{6.2}$$

We know that $K(p,0)|_{m=0}$ is finite and we see from (2.6) that

$$m\frac{\partial S(p')}{\partial m}\bigg|_{m\sim 0}=C(2\epsilon+1)\frac{m}{p^2}\left(\frac{m^2}{p^2}\right)^\epsilon a_0 \tag{6.3}$$

when $\epsilon>0$ [and as $m\to 0$, no divergent integral which multiplies m appears in (6.2)]. Thus (6.2) becomes

$$\delta m=m\,, \tag{6.4}$$

i.e., the electromagnetic mass δm, when expressed in terms of the physical mass m, is identically equal to it for all values of m.[21]

We can gain some insight into (6.4) by looking at the subtracted equation (2.13) for $\bar{m}(p)$. In (2.13) the quantity p_0 acts as a cutoff for the integrals generated by the perturbation expansion of $\Sigma^*(p)-\Sigma^*(p_0)$. Equation (2.13) can then be interpreted as the equation for the electron propagator in a theory which contains a cutoff p_0 and in which the mechanical mass is $\bar{m}(p_0)$. For values of the cutoff, $p_0\gg m$, the mechanical mass $\bar{m}(p_0)$ is then determined in terms of the physical mass m and the cutoff p_0 by (2.32); i.e.,

$$\bar{m}(p_0)=a_0(e_0^2)m(m^2/p_0^2)^\epsilon\,, \tag{6.5}$$

which is the sum of the usual perturbation-theory logarithms in the expansion of the bare mass in terms of the physical mass and the cutoff. From (6.5) we see that as the cutoff becomes larger it requires less and less mechanical mass to generate the same physical mass and in the limit when the cutoff p_0 becomes infinite, it takes only a vanishingly small mechanical mass to generate a finite physical mass.[21] Accordingly, in this limit δm becomes equal to m.

VII. CONSERVATION LAWS

The axial-vector current $j_5^\mu(x)=-i\bar{\psi}(x)\gamma^\mu\gamma_5\psi(x)$ obeys the formal equation of motion,

$$\partial_\mu j_5^\mu=2m_0\bar{\psi}\gamma_5\psi=2m_0 j_5\,, \tag{7.1}$$

where m_0 is the mechanical mass.[22] We wish to discuss (7.1) in the context of the renormalized theory. The unrenormalized operator $j_5^\mu(x)$ exists because the divergent part of every diagram for the proper vertex is the same as that of the corresponding proper vertex of the unrenormalized vector current $\bar{\psi}\gamma^\mu\psi$ which does exist. The same cannot be said for the unrenormalized operator $j_5(x)$. Hence, the argument that $m_0=0$ implies a conservation law for the axial current must be regarded with some caution, since when (7.1) is made precise with the use of a cutoff, both m_0 and $j_5(x)$ diverge in perturbation theory in the limit as the cutoff is removed, whereas the left-hand side of (7.1) remains finite. This difficulty was first pointed out by Maris and Jacob.[23]

[21] In deriving (6.4) we used (2.2), in which all internal photon propagators $D(k)$ in Σ^* were replaced by their asymptotic value e_0^2/k^2. It is easy to see, using scaling arguments, that the corrections (Ref. 5) to the asymptotic limit for $D(k)$ yield a contribution to Σ^* of the form $\gamma\cdot p(m^2/p^2)^{K(e_0^2)}+m(m^2/p^2)^{\epsilon+K(e_0^2)}$. If we then look at the high-p limit of the exact version of (2.2), we again find $\delta m=m$. Thus (6.4) is valid in the complete quantum electrodynamics including photon self-energy insertions provided $D(k^2)\to e_0^2/k^2$ as $k^2\to\infty$.

[22] We may remark also that anomalies in the divergence of the axial current of the sort discussed by S. Adler, Phys. Rev. **177**, 2426 (1969), R. Jackiw and K. Johnson, *ibid.* **182**, 1459 (1969), and others play no role because we can use a non-gauge-invariant axial-vector current whose divergence is consistent with (7.1) to discuss the Goldstone phenomena formally.

[23] Th. A. J. Maris and G. Jacob, Phys. Rev. Letters **17**, 1300 (1966).

To discuss this limit we shall merely paraphrase an argument already given.[24] A general matrix element of (7.1) can be obtained by the symmetrical insertion into single lines of the off-shell version of (7.1). Therefore, consider the equation for the proper vertex,

$$q^\mu\Gamma_\mu{}^5(p+q,p)=S^{-1}(p+q)\gamma_5+\gamma_5S^{-1}(p)+2m_0\Gamma^5(p+q,p)\,. \quad (7.2)$$

Here $\Gamma_\mu{}^5$ is the proper vertex corresponding to the axial-vector current and Γ^5 is the proper vertex corresponding to the pseudoscalar density $j_5(x)$.

We may re-express (7.2) in terms of renormalized quantities in the form

$$(Z_2/Z_1{}^A)q^\mu\bar{\Gamma}_\mu{}^S(p+q,p)=\bar{S}^{-1}(p+q)\gamma_5+\gamma_5\bar{S}^{-1}(p)+2m(Z_1{}^S/Z_1{}^{PS})\bar{\Gamma}^5(p+q,p)\,, \quad (7.3)$$

where

$$m_0=mZ_1{}^S/Z_2\,, \quad (7.4)$$

$Z_1{}^A$ is a renormalization constant of the axial-vector vertex, $Z_1{}^S$ is a suitably defined renormalization constant of the proper scalar vertex, and $Z_1{}^{PS}$ is the corresponding quantity for the pseudoscalar vertex. In the limit as $\Lambda\to\infty$, $\bar{\Gamma}_\mu{}^5$, $\bar{S}$, $\bar{\Gamma}^5$, and $Z_2/Z_1{}^A$ exist order by order in perturbation theory, so that

$$\frac{Z_1{}^S}{Z_1{}^{PS}}=\frac{Z_1{}^S/Z_2}{Z_1{}^{PS}/Z_2}=\frac{Z^S}{Z^{PS}}$$

approaches a finite nonvanishing limit, order by order, which is the ratio of the renormalization constants of the scalar and pseudoscalar currents. Since when we sum up the perturbation-theory contributions, $m_0=0$, and Z_2 is finite, Z^S vanishes as the cutoff tends to infinity. Since the ratio Z^S/Z^{PS} is finite (in our gauge), $Z_1{}^{PS}$ also vanishes in this limit [which we could have demonstrated by treating this vertex in exactly the same fashion as we discussed $m(p_0)$ in the earlier sections of this paper]. We can see the reason intuitively by observing that the divergent parts of the scalar and pseudoscalar vertices are independent of mass terms, and hence are the same since the interaction vertex is chirally invariant.

In conclusion, we find that in spite of a vanishing mechanical mass, there is no conservation of the unrenormalized axial-vector current and hence no chiral symmetry which is broken by the finite physical mass (at least in the sense of the sort of symmetry which when broken is accompanied by a Goldstone boson).

VIII. CONCLUSION

We have shown that if the photon propagator is set equal to $1/k^2$, then the high-p^2 behavior of the sum of all the terms in the perturbation expansion of $S(p)$ is given by (2.6). The coefficient in the power-series expansion of ϵ are determined by the renormalized perturbation expansion of the $m=0$, electron, positron, Bethe-Salpeter kernel K. This exact $(m^2/p^2)^\epsilon$ behavior leads to finite self-energy integrals, giving $\delta m\equiv m$ for all values of m.

To complete this study of the short-distance behavior of quantum electrodynamics, we must calculate the function $f(\alpha_0)$,[2] in order to determine whether the basic assumption, $e^2D(k)\to e_0{}^2/k^2$, $e_0{}^2<\infty$, is justified.

[24] G. Preparata and W. I. Weisberger, Phys. Rev. 175, 1965 (1968).

Appendix C. "Some Speculations on High-Energy Quantum Electrodynamics"

This paper is K. Johnson and M. Baker, Phys. Rev. **D8**, 1110 (1973). Reprinted with the kind permission of Physical Review D.

PHYSICAL REVIEW D VOLUME 8, NUMBER 4 15 AUGUST 1973

Some Speculations on High-Energy Quantum Electrodynamics*

K. Johnson†
University of Washington, Seattle, Washington 98195
and Massachusetts Institute of Technology, Cambridge, Massachusetts 02139‡

M. Baker
University of Washington, Seattle, Washington 98195
(Received 3 May 1973)

Recent work on quantum electrodynamics is reviewed, some speculations about the theory are made, and some conceivable future experimental implications are discussed.

I. INTRODUCTION

In this note we will first summarize the reasoning which leads to the possibility of a finite quantum electrodynamics. The arguments presented will in essence be those already discussed in previous publications.[1-7] However, we hope to be able to present the argument in a form which makes clear the beautiful simplicity of quantum electrodynamics. We will then discuss the implications of a finite theory for high-energy electrodynamic experiments.

In Sec. II we will show that a self-consistent finite solution of the quantum electrodynamics of zero-*physical*-mass electrons exists if the square[8] of the coupling constant x is chosen to be a positive root x_0 of the equation

$$f(x_0)=0\,, \tag{1.1}$$

where $(x/2\pi)f(x)$ is the sum of the coefficients of the logarithmically divergent integrals for the vacuum polarization in massless electrodynamics with coupling constant x.

In Sec. III we assume the existence of a root x_0 of Eq. (1.1) and study the properties of massless electrodynamics with coupling constant x_0, which is a self-consistent finite theory. We show that in this theory scattering amplitudes involving only external photons vanish and we obtain the following simplified equation for x_0:

$$f_1(x_0)=0\,, \tag{1.2}$$

where $f_1(x)$ is the contribution to $f(x)$ arising from diagrams containing a single closed fermion loop. Examples of diagrams which contribute to $f_1(x)$ are depicted in Fig. 1.

In Sec. IV we show that the quantum electrodynamics of electrons with physical mass $m\neq 0$ can be finite under two possible circumstances:

(i) The bare fine-structure constant α_0 is taken equal to x_0 [α_0 is determined from the high-momentum behavior of the photon propagator $D_{\mu\nu}(k)$].

(ii) The physical fine-structure constant α is taken equal to x_0 [α is determined from the behavior of $D_{\mu\nu}(k)$ for k^2 near zero].

If $m=0$, since no other scale must be introduced into a finite theory, the exact photon propagator $D(k)$ is proportional to the free propagator and hence the bare charge and the physical charge are equal. Alternatives (i) and (ii) are then equivalent and reduce to the result of Secs. II and III for massless electrodynamics. In the real world where $m\neq 0$ the distinct possibilities (i) and (ii) arise from two different orders of summing the series for the contributions to the vacuum polarization which depend upon the mass m. The order of summation which yields alternative (ii) was pointed out in the important work[7] of Adler. It yields the physically attractive possibility of determining theoretically the observed fine-structure constant α.

In Sec. V we review what is presently known about the fundamental function $f_1(x)$ and discuss the possibility of distinguishing between alternatives (i) and (ii) from high-energy experiments.

II. $m=0$ ELECTRODYNAMICS

The photon propagator $D_{\mu\nu}(k)$ can be written as

$$D_{\mu\nu}(k)=\left(g_{\mu\nu}-\frac{k_\mu k_\nu}{k^2}\right)D(k^2)+b\,\frac{k_\mu k_\nu}{k^2}\,, \tag{2.1}$$

where b is a free parameter which determines the gauge. $D(k^2)$ is then determined in terms of the vacuum-polarization function $\Pi_{\mu\nu}(k)$ by the equations

$$D^{-1}(k^2)=k^2[1+\Pi(k^2)]\,, \tag{2.2}$$

$$\Pi_{\mu\nu}(k)=(k^2 g_{\mu\nu}-k_\mu k_\nu)\Pi(k^2)\,. \tag{2.3}$$

The electron propagator $S(p)$ is determined in terms of the electron self-energy function $\Sigma(p)$ by the equation

$$S^{-1}(\gamma\cdot p)=\gamma\cdot p+\Sigma(p)\,. \tag{2.4}$$

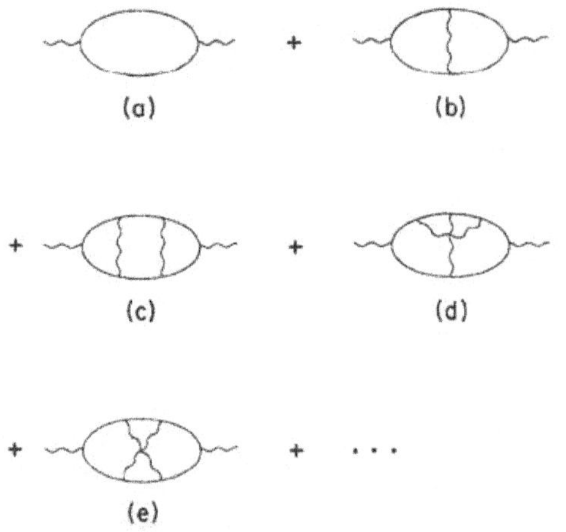

FIG. 1. Examples of diagrams contributing to $f_1(x)$.

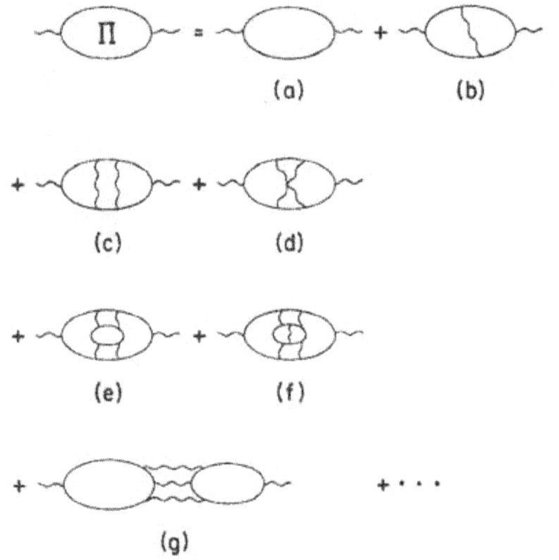

FIG. 3. Examples of graphs for $\Pi(k^2)$.

$\Sigma(p)$ and $\Pi_{\mu\nu}(k)$ are given by a sum over all Feynman graphs, examples of which are depicted in Figs. 2 and 3. In these diagrams the solid lines stand for the electron propagator $S(p)$ and the wavy lines for the photon propagator $D_{\mu\nu}(k)$. If a finite, unique solution to these equations exists, then by scale invariance

$$D(k)=1/k^2, \tag{2.5}$$

$$S(p)=1/\gamma\cdot p. \tag{2.6}$$

For simplicity of presentation we have omitted from Eqs. (2.5) and (2.6) nonessential finite constant factors which rescale the coupling constants appearing at each of the vertices in Figs. 2 and 3. Let us call the square[8] of this rescaled coupling constant x.

We now insert the trial solutions (2.5) and (2.6) for S and D into the expansions for Σ and Π depicted in Figs. 2 and 3. This substitution gives rise to all the perturbation-theory integrals for Π and Σ in massless electrodynamics except for those containing internal electron or photon self-

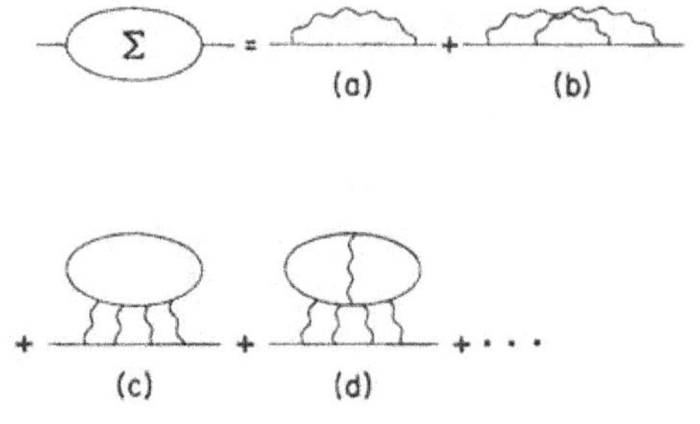

FIG. 2. Examples of graphs for $\Sigma(p)$.

energy corrections. Because $m=0$, the integrals not only possess the usual ultraviolet divergences, but they might also diverge in the "infrared" region where some subset of the momenta p_i of the internal lines become small. However, in the Appendixes of Ref. 3 it was shown by elementary power-counting arguments that no such "infrared divergence" arises when one or more of the p_i are held fixed. Furthermore, as long as the external electron and photon momenta are nonvanishing, there are no divergences arising from the small-p_i integration region when all the p_i are integrated over. This insensitivity of the relevant integrals of quantum electrodynamics to the electron mass m when m is set equal to zero is the essential feature of quantum electrodynamics which leads to the possibility of a consistent finite theory.

The situations concerning ultraviolet divergences in the above expressions for Σ and Π are quite distinct. For Σ the usual ultraviolet divergences in the perturbation-theory integrals can be isolated in terms of two infinite constants, the wave-function renormalization constant Z_2 and the electron self-mass δm. However, since δm is proportional to m and since m has been set equal to zero, the δm divergence is not present. The expansion of Fig. 2 then contains only the Z_2 divergence. However, this divergence depends upon the value of the gauge parameter b and can be eliminated in every order of perturbation theory with a suitable choice of b. The proof of this fact makes essential use of the properties of Feynman integrals for small as well as large values of the momenta of the external lines. In a suitably chosen gauge the integrals of Fig. 2 for Σ possess neither infrared[9] nor ultraviolet divergences. By dimensional arguments it follows that Σ is proportional to $\gamma \cdot p$ multiplied by a power series in x with coefficients which are finite to every order in perturbation theory. Thus (2.5) and (2.6) are self-consistent solutions of Eq. (2.4). Equation (2.4) then simply determines the constant factor we have omitted from the expression (2.6) for S in terms of the coupling constant x.

We now investigate the ultraviolet divergences in Π to see if (2.5) and (2.6) can also be self-consistent solutions of Eq. (2.2). There is an over-all superficial quadratic divergence in the perturbation-theory integrals of Fig. 3. However, because of the factor $k^2 g_{\mu\nu} - k_\mu k_\nu$ which appears in the expression (2.3) for $\Pi_{\mu\nu}$, this quadratic divergence is reduced to the usual logarithmic ultraviolet divergence for Π. This over-all logarithmic divergence arises from the final integration over the momentum p of the electron line which is coupled to the external photon line on the left of the diagrams of Fig. 3. If p is held fixed, the integrations over the remaining internal lines in these diagrams are finite in the ultraviolet region. This follows from the following facts: (a) The diagrams of Fig. 3 contain neither electron nor photon self-energy insertions; (b) the only other kind of insertions which lead to ultraviolet-divergent subintegrations are vertex insertions as in Fig. 3(d), which, because of Ward's identity, are finite in the gauge in which Σ is finite.[10]

Using scale invariance we thus conclude that the integrals for $\Pi(k)$ can be written in the form

$$\Pi(k^2, x) = \frac{x}{2\pi} \int_0^\infty \frac{dp^2}{p^2} f(p^2/k^2, x), \tag{2.7}$$

where $f(p^2/k^2, x)$ is a finite function of p^2/k^2 and x which arises from carrying out all the integrations in the diagrams of Fig. 3 except the final integration over the magnitude of p. Equation (2.7) follows from the fact that the integrals for f are both infrared- and ultraviolet-finite. The ultraviolet divergence in Π is obtained by calculating the large-p limit of f. But by scale invariance this is equivalent to letting k approach zero. Now, since, as stated above, all integrals for f remain infrared-finite when k is set equal to zero, we conclude

$$\lim_{p^2 \to \infty} f(p^2/k^2, x) = f(x), \tag{2.8}$$

where $f(x)$ is the finite function of the coupling constant x which is obtained by setting $k = 0$ in the integrals for $f(p^2/k^2, x)$.[11] We can then write

$$\begin{aligned}\Pi(k^2, x) = {} & \frac{x}{2\pi} \int_0^{k^2} \frac{dp^2}{p^2} f(p^2/k^2, x) \\ & + \frac{x}{2\pi} \int_{k^2}^\infty \frac{dp^2}{p^2} [f(p^2/k^2, x) - f(\infty, x)] \\ & + \frac{x}{2\pi} f(x) \int_{k^2}^\infty \frac{dp^2}{p^2}.\end{aligned} \tag{2.9}$$

We see that the Π divergence is like a single power of the logarithm of an ultraviolet cutoff on the momentum p^2, and therefore the assumption of the existence of a finite solution to mass-zero electrodynamics is in general inconsistent. However, if the coupling constant x is chosen to be a root x_0 of the equation

$$f(x) = 0, \tag{2.10}$$

then the coefficient of this logarithmic divergence vanishes.[12] $\Pi(k^2, x_0)$ is then finite and by scale invariance [or by explicitly letting $p^2 \to k^2 z$ in the finite terms in (2.9)] is a constant $\Pi(x_0)$ independent of k^2. Thus, for coupling constant $x = x_0$, (2.5) and (2.6) are also self-consistent solutions of Eq. (2.2). Equation (2.2) then simply determines the constant factor we have omitted from Eq. (2.5) for D in terms of the finite constant $\Pi(x_0)$. We thus conclude that mass-zero quantum electrodynamics with coupling constant x_0 determined from Eq. (2.10) is a finite theory having the exact electron and photon propagators which are proportional to the free propagators.

The possibility of obtaining a finite mass-zero theory by imposing a single condition upon x depended upon the fact that the integrals of Fig. 3 for Π diverged only like a single power of a logarithm. The essential reason for this single logarithm was that all integrations except the final integration over p were finite in both the ultraviolet and infrared regions. If the integrals defining $f(p^2/k^2)$ of Eq. (2.7) possessed infrared divergences when $k = 0$, then f could have contained terms like $[\ln(p^2/k^2)]^n\ n \geq 1$ and the above argument leading to a finite theory could not have been carried out.

The arguments of this section leading to the possibility of a finite mass-zero electrodynamics depended only upon the explicit properties of perturbation-theory integrals. In Sec. III we will examine the properties of this finite theory using some results which follow from the operator representations of scattering amplitudes in field theory. Such results are not explicitly visible from the perturbation solution for these amplitudes and hence the theoretical basis for the simplifications to be obtained in Sec. III is fundamentally different from that for the results obtained above.

III. PROPERTIES OF MASS-ZERO ELECTRODYNAMICS

We now assume there exists a positive root x_0 of Eq. (2.10). Then the quantum electrodynamics of photons interacting with massless electrons with coupling constant x_0 is a finite relativistic quantum field theory. In this theory the exact photon propagator $D(k)$ is proportional to the free propagator, and hence the absorptive part of $D(k)$ vanishes. However, the absorptive part of $D(k)$ is determined by the Fourier transform of $\langle 0|j_\mu(x)j_\nu(y)|0\rangle$, where $j_\mu(x)$ is the electromagnetic current operator. We then conclude that the current operator in mass-zero electrodynamics satisfies the equation

$$\langle 0|j_\mu(x)j_\nu(y)|0\rangle = 0. \tag{3.1}$$

But we now apply the theorem[13]

$$\langle 0|j_\mu(x)j_\nu(y)|0\rangle = 0 \text{ implies}$$

$$\langle 0|j_{\mu_1}(x_1)\cdots j_{\mu_n}(x_n)|0\rangle = 0. \tag{3.2}$$

Thus all amplitudes involving only external photon lines (real or virtual) vanish. The result (3.2), unlike the previous ones of our discussion, cannot be understood from an examination of the properties of perturbation theory. That is, in perturbation theory an n-photon amplitude is expressed as a sum of all closed-loop diagrams with n external photon lines. Of course, to a given order in perturbation theory these diagrams do not vanish. The mechanism by which the sum over all diagrams vanishes is not at all clear. However, this vanishing must occur if one accepts the general principles upon which it is based.

If this theorem[13] could be extended to include processes where there are external electron lines, then all transition amplitudes would vanish. This would mean that the finite quantum electrodynamics of zero-mass electrons would be equivalent to a free-field theory. We would thus conclude that in the absence of an electron mass there are no interactions, i.e., interactions are a consequence of the nonvanishing electron mass. The difficulty of this extension is caused by the absence of a positive metric in the portion of the Hilbert space associated with external charged lines. However, in this paper we will not make any use of this speculation about the zero-mass theory.

We now show that the vanishing Eq. (3.2) of the n-photon amplitudes implies that x_0 satisfies the simpler Eq. (1.2). The vanishing of the four-photon amplitude means that the sum of all graphs of the type depicted in Fig. 4 which involve an internal fully interacting four-photon amplitude must vanish. Likewise the sum of all diagrams of the class of Fig. 5 vanishes because the multiphoton

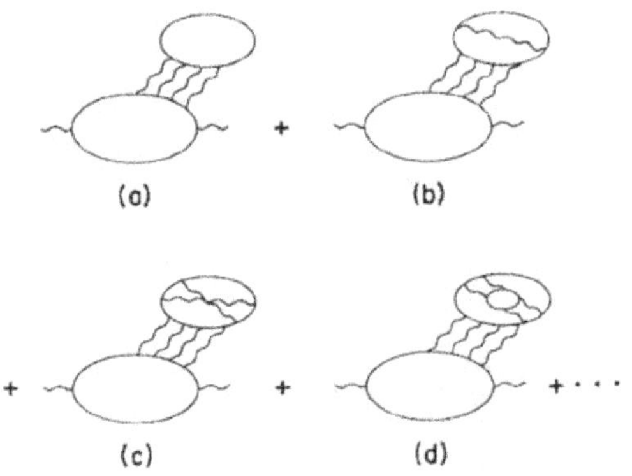

FIG. 4. Examples of contributions to $f(x)$ from graphs involving interacting photons.

interaction on the right vanishes when all the diagrams for that interaction are included. In general the diagrams for $f(x)$ break up into subclasses, each of which involves some subset of photons fully interacting with each other. Hence each of these subclasses must vanish separately. There remains the class of diagrams which involve no photon-photon interactions, namely those diagrams involving a single closed fermion loop, depicted in Fig. 1. These graphs give the contribution $f_1(x)$ to $f(x)$. Since all other contributions to $f(x)$ vanish at $x=x_0$, we see that $f_1(x)$ must also vanish at $x=x_0$.

If we apply the above argument to the n-photon amplitude, we immediately conclude that the single-closed-fermion-loop contribution to the n-photon amplitude vanishes by itself.

Adler[7] has made the important observation that the vanishing [(3.2)] of the n-photon amplitudes implies that all derivatives of $f(x)$ and $f_1(x)$ vanish at $x=x_0$. Another way of obtaining this result is to use the fact that the nth derivative of $f(x)$ is related to the forward n-photon$-$ n-photon amplitude. The latter can be seen by differentiating Eq. (2.9) with respect to x. This yields the relation[3]

$$\frac{d}{dx}f(x) = q^2\tfrac{1}{2}[K(0;q)], \tag{3.3}$$

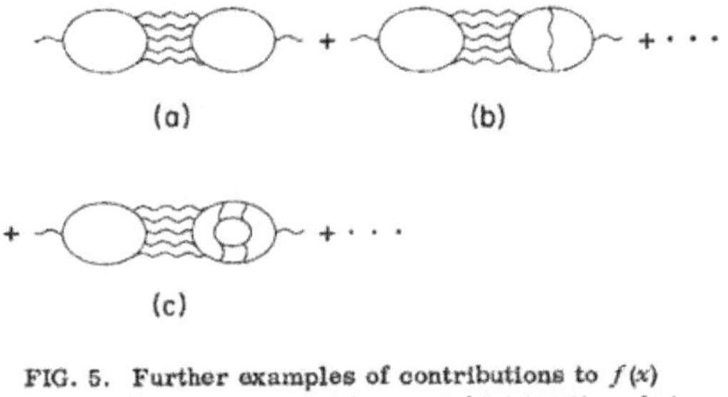

FIG. 5. Further examples of contributions to $f(x)$ from graphs containing a subgroup of interacting photons.

where $K(k; q)$ is the asymptotic Bethe-Salpeter kernel for the forward scattering of photons of momenta q and k. However, since when $x = x_0$ the scattering amplitude T for this process vanishes, so must the kernel K, since T and K are related by the usual linear scattering integral equation. Similarly differentiation of Eq. (3.3) with respect to x relates d^2f/dx^2 to the kernel for the forward 3→3 photon amplitude. Likewise differentiating Eq. (2.9) n times with respect to x relates $d^nf(x)/dx^n$ to the kernel for the forward $n \to n$ photon amplitude which vanishes at $x = x_0$. We then conclude

$$\frac{d^nf(x)}{dx^n} = 0 \quad \text{at } x = x_0. \tag{3.4}$$

The above reasoning also relates the nth derivative of the single-closed-loop function $f_1(x)$ to the single-closed-loop contribution to the forward n-photon→n-photon amplitude which vanishes at $x = x_0$. As a consequence, we obtain Adler's result,

$$\frac{d^n}{dx^n} f_1(x) = 0 \quad \text{at } x = x_0. \tag{3.5}$$

Thus both the functions $f(x)$ and $f_1(x)$ have essential singularities at $x = x_0$. Hence in order to test for the existence of a finite theory of mass-zero quantum electrodynamics, we must calculate the single-closed-loop function $f_1(x)$ and look for a value of the coupling for which $f_1(x)$ and all its derivatives vanish.

We would like to emphasize that the simplified eigenvalue Eq. (1.2) and the essential singularities of $f(x)$ and $f_1(x)$ are basically consequences of the theorem[13] of Eq. (3.2).

IV. QUANTUM ELECTRODYNAMICS OF FINITE-MASS ELECTRONS

We now turn to the theory of physical interest, ordinary quantum electrodynamics of electrons with mass $m \neq 0$. This theory will be consistent if the renormalized electron propagator $\bar{S}(p)$ and the renormalized photon propagator $\bar{D}(k)$ behave like free propagators at high energy, i.e.,

$$\lim_{p \to \infty} \bar{S}(p) \sim \text{const}/\gamma \cdot p, \tag{4.1}$$

$$\lim_{k^2 \to \infty} \alpha \bar{D}(k) \sim \alpha_0/k^2, \tag{4.2}$$

where α is the fine-structure constant and α_0 by definition is the bare fine-structure constant. We will see that the high-energy behavior of this theory is determined completely by the properties of mass-zero electrodynamics.

We first study the perturbation expansion for $\bar{D}(k)$ which is generated from the vacuum-polarization function $\bar{\Pi}(k)$ by the equations

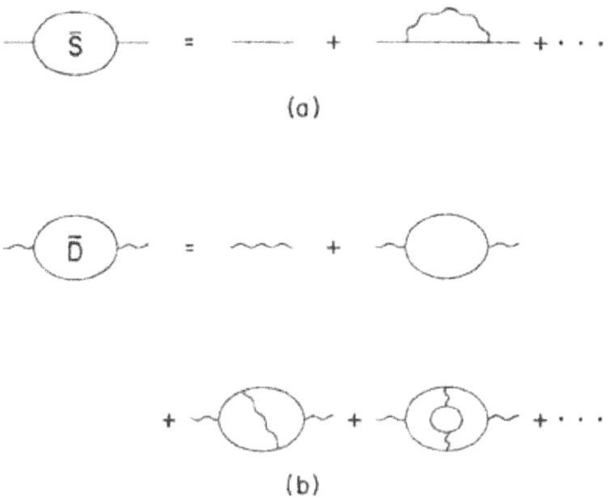

FIG. 6. Graphs representing the perturbation expansions of $\bar{S}(p)$ and $\bar{D}(k)$.

$$\bar{D}^{-1}(k) = k^2[1 + \alpha \bar{\Pi}_R(k)], \tag{4.3}$$

where

$$\bar{\Pi}_R(k) = \bar{\Pi}(k) - \bar{\Pi}(0). \tag{4.4}$$

For convenience we have removed from the definition of $\bar{\Pi}$ the factor α which may be associated with the coupling to the external photon lines. The perturbation expansion for $\bar{\Pi}(k)$ can be obtained from the sum of all diagrams of the type of Fig. 3 by replacing all electron and photon lines in each diagram by the renormalized electron and photon propagators $\bar{S}(p)$ and $\bar{D}(k)$ of Fig. 6, and by associating with each internal vertex the renormalized charge e. The electron lines on the right-hand side of Fig. 6 of course stand for $1/(\gamma \cdot p + m)$, where m is the physical mass of the electron. The resulting expression for $\bar{\Pi}_R(k)$, Eq. (4.4), is the usual series of convergent renormalized perturbation-theory integrals, which for $k^2/m^2 \gg 1$ takes on the form

$$\bar{\Pi}_R(k) = c_0(\alpha) + c_1(\alpha) \ln \frac{k^2}{m^2} + \sum_{n=2}^{\infty} c_n(\alpha) \left(\ln \frac{k^2}{m^2} \right)^n + O\left(\frac{m^2}{k^2} \right). \tag{4.5}$$

The coefficients $c_n(\alpha)$ are power series in the fine-structure constant α, obtained by evaluating the asymptotic form of the perturbation-theory integrals. The requirement (4.2) means that $\bar{\Pi}_R(k)$ must remain finite as $k^2 \to \infty$. This condition is equivalent to the requirement that the unrenormalized theory is finite or, equivalently, that Z_3 is finite.

We must thus evaluate the high-k^2 limit of $\bar{\Pi}_R(k)$. We first do this according to the procedure of Ref. 3. We make the following assumption (later to be checked by self-consistency):

(A) We can replace $\bar{\alpha}D(k)$ by its limiting form α_0/k^2 in the expression for $\bar{\Pi}(k)$ in order to calculate the dominant large-k behavior of $\bar{\Pi}_R(k)$. That is, we assume that the leading corrections to the asymptotic limit (4.2) fall off sufficiently rapidly that they yield contributions to $\bar{\Pi}(k)$ which will vanish as $k\to\infty$. Furthermore, if we choose the gauge in which Z_2 is finite, the electron self-energy corrections are unimportant. Thus with assumption (A) the large-k behavior of $\bar{\Pi}(k)$ is obtained from the sum of the diagrams of Fig. 3 by replacing each internal photon line by $1/k^2$, each vertex by $\sqrt{\alpha_0}$, and each internal electron line by $1/(\gamma\cdot p + m)$. We are then left with a series for the asymptotic behavior of $\bar{\Pi}(k)$ which is the same as the series for $\Pi(k)$ in the mass-zero theory except for the presence of the electron mass m in the denominators.

Since $\bar{\Pi}_R(k)$ is a function of k^2/m^2, we can calculate the $\lim_{k\to\infty}\bar{\Pi}(k)$ by taking the limit $m\to 0$. By our previous discussion of Eqs. (2.7)–(2.9), we know that all integrals for $\bar{\Pi}_R(k)$ are finite when m is set equal to zero except for the final integration over p, which with $m=0$ would be infrared-divergent because of $\bar{\Pi}(0)$ in Eq. (4.4). The presence of the electron mass in this final integration cuts off this divergence, and the subtraction in (4.4) makes the final integration over p ultraviolet-convergent. Thus we can set $m=0$ everywhere except in the final integration over p. This effectively replaces the p^2 in the denominators of the integrals in Eqs. (2.7) and (2.9) by p^2+m^2. Then, using (2.9) and (2.4), we conclude that

$$\lim_{k\to\infty}\bar{\Pi}_R\left(\frac{k^2}{m^2}\right)=\lim_{m\to 0}\bar{\Pi}_R\left(\frac{k^2}{m^2}\right) \sim -\frac{1}{2\pi}f(\alpha_0)\ln\frac{k^2}{m^2}+\text{constant}. \tag{4.6}$$

We can understand the relation between Eqs. (4.5) and (4.6) by first noting that from the previous discussion it follows that the higher powers of $\ln(k^2/m^2)$ in Eq. (4.5) come only from diagrams which include the vacuum-polarization corrections of Fig. 6(b) in the internal photon lines of Fig. 3. Thus, for example, when the internal photon line in Fig. 3(b) is replaced in the diagrams of Fig. 6(b) for $\bar{D}$, we obtain the diagrams depicted in Fig. 7, which individually behave asymptotically like high powers of $\ln(k^2/m^2)$. However, if we first sum over all diagrams of Fig. 7, then from (A) it follows that the high-k^2 limit of these diagrams is the same as that of Fig. 3(b) with α replaced by α_0. Thus if we take the $\lim k\to\infty$ of the individual terms in Fig. 7 we generate a series of the structure (4.5), while if we first sum over all diagrams

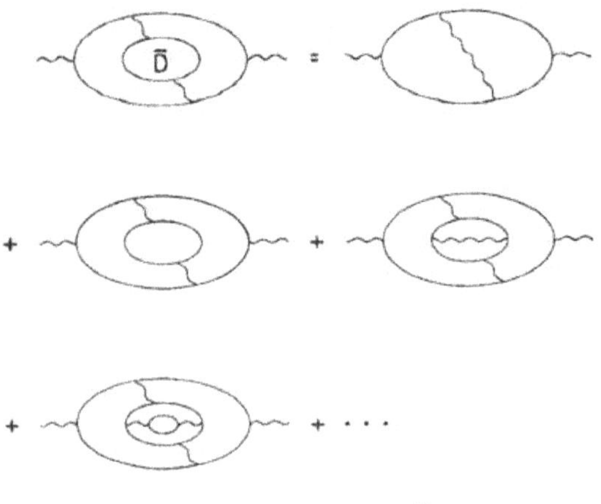

FIG. 7. Examples of contributions to $\bar{\Pi}_R(k)$ from diagrams having vacuum-polarization insertions in internal photon lines.

and then go to the limit $k\to\infty$ we obtain a result proportional to $f(\alpha_0)\ln(k^2/m^2)$. There is a similar relation between the full series (4.5) and the result (4.6).

From (4.6) we conclude that under assumption (A) Eq. (4.2) can be self-consistent only if the bare charge α_0 is chosen to be a root x_0 of Eq. (2.10). If we choose $\alpha_0=x_0$, we find

$$\lim_{k^2\to\infty}\bar{\Pi}_R(k^2)=\text{constant} \tag{4.7}$$

for $\alpha_0=x_0$, and Eqs. (4.2) and (4.3) are consistent. It remains to verify (A), namely, to show that (4.7) is still valid when the corrections to the limit (4.2) are included in the calculation of $\bar{\Pi}_R(k)$. This was done in Ref. 3 using the incorrect assumption that $f'(x_0)\neq 0$. However, Adler[7] showed that this conclusion is still valid by using the fact [Eq. (3.2)] that the n-photon amplitude vanishes. We hence conclude that there exists a self-consistent solution for the photon propagator which behaves for large k^2 like the free propagator provided the bare coupling constant α_0 is chosen to be a root of Eq. (2.10). Furthermore, it can be readily seen[1,3] that the physical fine-structure constant α is left undetermined except that it must be less than $\alpha_0=x_0$. This latter fact is more easily seen from the Gell-Mann–Low equation,[1] which we will use in Sec. V of this paper in order to determine the corrections to Eq. (4.7) at finite values of k^2. Finally, we note that $\alpha_0=x_0$ is of course also a root of the simpler equation (1.2).

Adler[7] has pointed out that a different order of performing the summation over the diagrams for $\bar{\Pi}_R(k)$ and taking the limit $k^2\to\infty$ can lead to the conclusion that a consistent solution for $\bar{D}$ exists when the physical fine-structure constant α is chosen equal to x_0. The first step in Adler's procedure is to sum all diagrams in the renormalized perturbation expansion for $\bar{\Pi}_R(k)$ which contain a single fermion closed loop. There is no prior summation of vacuum-polarization insertions in internal photon lines. This series can thus be represented by the graphs of Fig. 1 where each vertex corresponds to the renormalized charge e and each internal photon line corresponds to free photon propagator $1/k^2$. From our previous discussion, it is clear that this series of diagrams gives a contribution to $\bar{\Pi}_R(k)$ which for $k^2 \to \infty$ behaves like

$$\bar{\Pi}_R(k) \underset{k^2\to\infty}{\sim} -\frac{1}{2\pi} f_1(\alpha) \ln\frac{k^2}{m^2} + \text{constant} + O\left(\frac{m^2}{k^2}\right). \tag{4.8}$$

Thus this subset of diagrams yields an asymptotically finite contribution to $\bar{\Pi}_R(k)$ if we choose $\alpha = x_0$, i.e., if α satisfies the equation

$$f_1(\alpha) = 0. \tag{4.9}$$

Adler assumes that $\alpha = x_0$ and considers contributions to $\bar{\Pi}_R(k)$ from diagrams containing two fermion closed loops, examples of which are depicted in Fig. 8. Since the individual diagrams of Fig. 8 contain vacuum-polarization insertions, they behave like higher powers of $\ln(k^2/m^2)$ for large k^2. However, the sum of all diagrams of the type in Fig. 8 contains an inner closed loop with all possible photon exchanges. Because of the eigenvalue condition [(4.9)], this gives a contribution to internal vacuum polarization in Fig. 8 which is asymptotically finite. The sum of the diagrams of Fig. 8 is thus equivalent to the diagram of Fig. 9. The crosses at the internal vertices of Fig. 9 indicate that the effective constant at these vertices differs from e because of the contribution of the finite part of the vacuum-polarization insertions in Fig. 8. By our previous argument, Fig. 9, and hence the sum of the diagrams of Fig. 8, then behave like a single power of $\ln(k^2/m^2)$.

Now consider the sum of diagrams of Fig. 10(a)

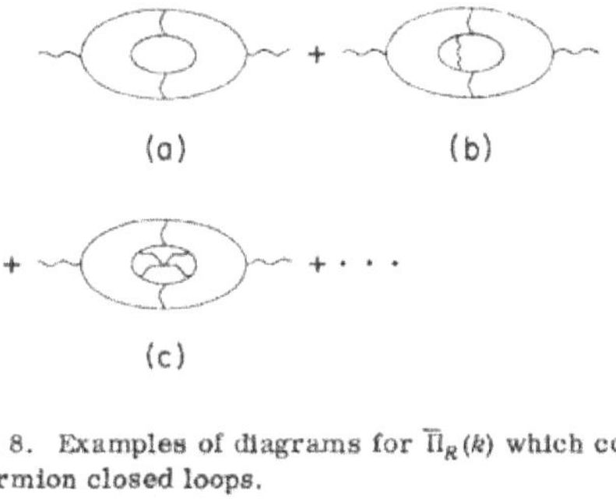

FIG. 8. Examples of diagrams for $\bar{\Pi}_R(k)$ which contain two fermion closed loops.

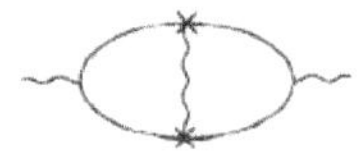

FIG. 9. Graph representing contribution of the sum of the graphs of Fig. 8.

where the outer closed loop contains an additional internal photon line. The sum of all diagrams of Fig. 10(a) is then equivalent to the diagram of Fig. 10(b). Likewise the sum over all diagrams which contain any number of internal photons in either closed fermion loop is equivalent to the sum of diagrams depicted in Fig. 11. From our previous discussion we see that each diagram of Fig. 11 behaves for large k^2 like a single power of $\ln(k^2/m^2)$ with a coefficient which is determined by setting $m=0$ and performing all integrations except one. Let us then carry out all integrations in each diagram of Fig. 11 except for the integration over the photon line which joins the two crosses. These integrations generate diagrams for the single-loop contribution to photon-photon scattering.[14] (See Fig. 12.) (The external lines for this photon-photon amplitude are the two external photons of the diagrams of Fig. 11 and the two crosses.) Thus the coefficient of $\ln(k^2/m^2)$ in the sum of the diagrams of Fig. 11 is proportional to the single-closed-loop contribution to photon-photon scattering in mass-zero electrodynamics with coupling constant α. However, since α satisfied Eq. (4.9), it follows from Eq. (3.2) that this photon-photon amplitude vanishes. Hence the coefficient of $\ln(k^2/m^2)$ in the sum of the diagrams of Fig. 11 vanishes. Thus the sum of the two-fermion-closed-loop diagrams represented by Fig. 11 gives

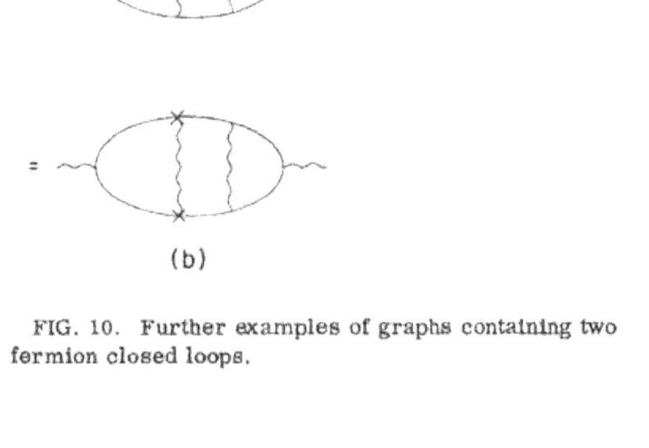

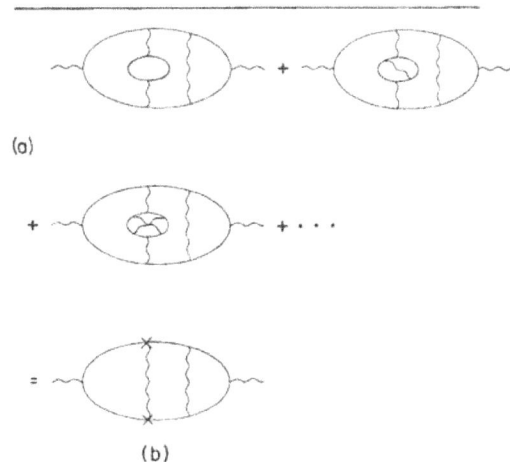

FIG. 10. Further examples of graphs containing two fermion closed loops.

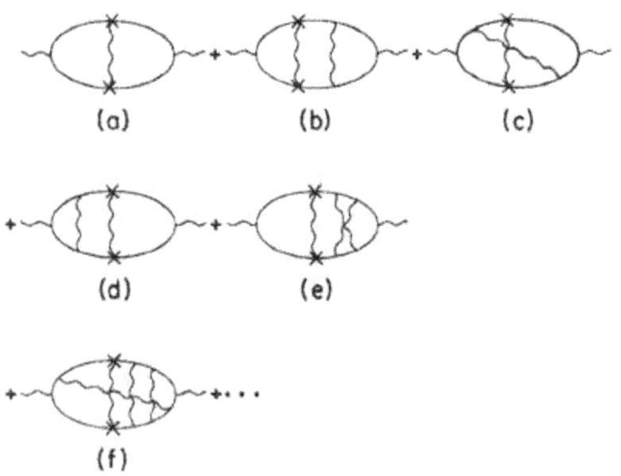

FIG. 11. Graphs representing sum of diagrams with two closed loops, each of which contains any number of internal photons.

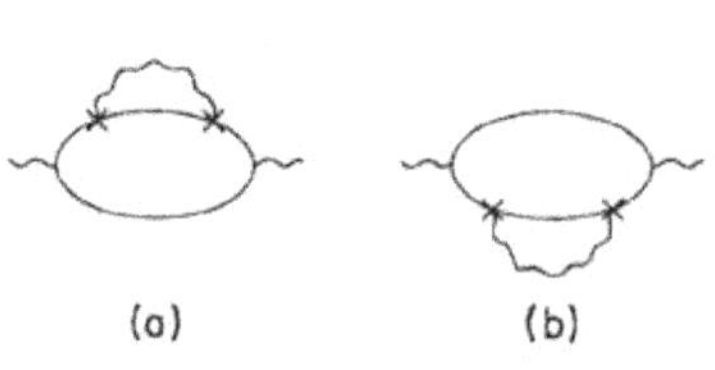

FIG. 12. Graphs representing two-closed-loop diagrams which contain an electron self-energy insertion.

a contribution to the vacuum polarization $\overline{\Pi}_R(k^2)$ which for $k^2/m^2 \gg 1$ has the behavior

$$\overline{\Pi}_R(k) \xrightarrow[k^2\to\infty]{} \text{constant} + O\left(\frac{m^2}{k^2}\right). \qquad (4.10)$$

Therefore if α satisfies (4.9), the sum over all diagrams in which the inner and outer fermion loops of Fig. 8(a) contain any number of internal lines gives a finite asymptotic contribution to the vacuum polarization.

The basic ingredients in the above argument are the following:

(a) diagrams for $\overline{\Pi}_R(k)$ without vacuum-polarization insertions behave like a single power of $\ln(k^2/m^2)$ for large k^2, and

(b) amplitudes with n external photons in mass-zero electrodynamics and coupling α vanish if $\alpha = x_0$.

Using (a) and (b) it is easy to see that if we sum over all diagrams containing a fixed number of closed fermion loops then $\Pi_R(k^2)$ will have the behavior (4.10) if α is chosen equal to x_0. In our previous vacuum-polarization insertion summation procedure based upon assumption (A), we summed over diagrams containing an infinite number of closed loops at the first stage when we replaced $\overline{D}$ by α_0/k^2 as in the example of Fig. 7. Thus in order to go from this procedure which leads to (4.10) and an eigenvalue equation for α, we must interchange the limit $k\to\infty$ with the summation over all diagrams.

If Adler's loopwise summation procedure for calculating $\overline{\Pi}(k)$ yields the physically correct solution of quantum electrodynamics, then the fine-structure constant α is fixed and equal to x_0. Now it follows from general arguments that the bare fine-structure constant α_0 is greater than α. Since the function $f_1(x)$ has an essential singularity at $x = x_0$, and $\alpha_0 > \alpha = x_0$, the point α_0 lies outside the radius of convergence of the function $f_1(x)$. Thus the rearrangement of diagrams and the interchange of the limit $k\to\infty$ with the summation over all diagrams which is necessary to convert the "loopwise" summation procedure into the vacuum-polarization insertion procedure [assumption (A)] is not justified. Hence one cannot deduce Eq. (4.6) and the eigenvalue condition for α_0. Conversely, suppose that the vacuum-polarization insertion assumption (A) yields the physically correct solution for $\overline{\Pi}_R(k)$; then $\alpha_0 = x_0$ and α is left undetermined except for the requirement that it be less than α_0. In this case the loopwise summation procedure will not work since $f_1(\alpha)$ does not vanish and hence the single-closed-loop diagram is not finite as $k^2 \to \infty$.

We now turn to the equation for the renormalized electron propagator $\overline{S}(p)$. We assume

(1) that there exists a consistent finite solution for $\overline{D}$ which has the behavior (4.2), and

(2) that we can replace $\alpha\overline{D}$ by its limiting form α_0/k^2 in the expression for the electron self-energy function $\overline{\Sigma}(p)$ in order to calculate the dominant high-energy behavior of $\overline{S}(p)$.

We know that assumption (1) is justified if there exists a positive root of the equation $f(x_0) = 0$. In the case $\alpha = x_0$, the asymptotic corrections to $\overline{\Pi}_R(k)$ [Eq. (4.10)] are of order m^2/k^2 and so assumption (2) is justified. In the case $\alpha_0 = x_0$, we will see in Sec. V that the asymptotic corrections to $\overline{\Pi}_R(k)$ vanish more slowly as $k^2 = 0$. In this case Adler has noted that assumption (2) will still be valid if the theorem of Eq. (3.2) can be extended to amplitudes containing external fermion lines.

Using (1) and (2) the Schwinger-Dyson equation for $\overline{S}$ then becomes an integral equation whose kernel is the Bethe-Salpeter kernel for electron-positron scattering. This equation was treated in detail in Ref. 5 and we will only state the results of that analysis here, since our main concern in this paper is the photon propagator. It was shown there that the solution of this equation for $\overline{S}(p)$ for $p^2 \gg m^2$ depended crucially upon the behavior of the kernel for $m=0$. Moreover, the kernel is sufficiently well behaved so that in a suitably chosen gauge the solution for $\bar{S}(p)$ takes on the form

$$\bar{S}(p) \xrightarrow[p^2/m^2 \gg 1]{} \text{constant} \times \left[\frac{1}{\gamma \cdot p} + \frac{m}{p^2} a \left(\frac{m^2}{p^2} \right)^{\epsilon} \right], \tag{4.11}$$

where a and ϵ are constants. The parameter ϵ is determined from the expansion of the kernel for the scattering of zero-mass electrons and positrons with coupling constant α_0. To order α_0^2, ϵ is given by[5]

$$\epsilon = \frac{3}{2}\frac{\alpha_0}{2\pi} + \frac{3}{8}\left(\frac{\alpha_0}{2\pi}\right)^2 + \cdots . \tag{4.12}$$

We see that Eq. (4.11) is consistent with (4.1) if $\epsilon > -\frac{1}{2}$. If $\epsilon < -\frac{1}{2}$, the mass term dominates the $1/\gamma \cdot p$ term and there are no consistent finite solutions for electron propagators.

If we introduce a cutoff Λ, then the bare mass $m_0(\Lambda)$ is given in terms of the physical mass m by the relation[5]

$$m_0(\Lambda) = am(m^2/\Lambda^2)^{\epsilon} . \tag{4.13}$$

Thus if $\epsilon > 0$, then $\delta m = m - m_0(\Lambda)$ is finite as the cutoff $\Lambda \to \infty$. The usual perturbation-theory divergence for δm arises from putting the second term in Eq. (4.11) with $\epsilon = 0$ into the equation for $\bar{\Sigma}$. This "divergence" is thus clearly not intrinsic to quantum electrodynamics but is the result of the inapplicability of perturbation theory.

Although ϵ is positive to order α_0^2, we have no general proof of positivity. We cannot rule out the possibility that ϵ might lie in the range $-\frac{1}{2} < \epsilon < 0$. In this case Eq. (4.11) would still be a valid asymptotic solution of the renormalized equation for $\bar{S}$, although from Eq. (4.13) we would find $\delta m = \infty$.[15]

V. EXPERIMENTAL IMPLICATIONS OF ALTERNATIVES (i) AND (ii) OF SEC. I

We have thus shown that quantum electrodynamics can be a consistent finite theory with electron and photon propagators which behave like free propagators at high energy, provided there exists a positive root x_0 of the equation $f(x_0) = 0$. $(x/2\pi) \times f(x)$ is the coefficient of the divergent logarithm in the sum of the $m=0$ diagrams depicted in Fig. 3. If $f(x_0)=0$, then $f_1(x_0)=0$, where $(x/2\pi)f_1(x)$ is the coefficient of the divergent logarithm in the sum of the simpler set of diagrams of Fig. 1. Furthermore $(d^n/dx^n)f(x) = (d^n/dx^n)f_1(x) = 0$ at $x = x_0$. Conversely, if we calculate $f_1(x)$ and find a positive value of x for which $f_1(x)$ and all its derivatives vanish, we can be fairly sure that $f(x)$ also vanishes at this point. However, this has not yet been proven.

There are then the following alternatives:

(i) A positive root x_0 exists and $\alpha_0 = x_0$.

(ii) A positive root x_0 exists and $\alpha = x_0$.

(iii) No positive root x_0 exists (in this case we have found no finite consistent solution of electrodynamics).

The best way to decide among the above alternatives is to calculate $f_1(x)$. We have been trying to find ways to calculate $f_1(x)$ exactly for the past five years, but no real progress has been made since Rosner's[16] sixth-order calculation which includes all the diagrams of Fig. 1. There has not even been a satisfactory explanation of the simplicity of Rosner's result[16]

$$f_1^{(6)}(x) = \frac{2}{3} + \left(\frac{x}{2\pi}\right) - \frac{1}{4}\left(\frac{x}{2\pi}\right)^2 . \tag{5.1}$$

Rosner carried out his calculation in the gauge in which Z_2 is finite. The sixth calculation of $f_1(x)$ in the Feynman gauge has been carried out by Brandt.[17] The importance of Brandt's calculation, aside from explicitly verifying the gauge invariance of $f_1(x)$, was to show that the calculation in the Feynman gauge possessed the following great simplicity: The sum of the nontrivial graphs (d) and (e) of Fig. 1, which involve crossed photon lines, gives no contribution to $f_1(x)$ when calculated in this gauge. Thus the sixth-order result (5.1) for $f_1(x)$ is obtained by evaluating in the Feynman gauge the uncrossed 2-photon diagram 1(c) and the internal electron self-energy corrections, both of which are trivial to calculate. However, the reason why Figs. 1(d) plus 1(e) do not contribute in the Feynman gauge has not been understood, and there has been no successful use of this simplicity to calculate the higher orders of $f_1(x)$.

Another class of attempts to calculate $f_1(x)$ have made use of coordinate-space methods.[18] Since $f_1(x)$ is calculated from a theory in which the electron mass m is equal to zero, the integrals for f_1 are all conformally invariant. We have attempted to exploit the conformal invariance in order to calculate f_1 to all orders. Although a little progress has been made, the method has been plagued with certain subtleties.[18] We will report on this work in a separate publication, hopefully to stimulate further attempts to overcome our present difficulties.

The fact that $f_1(x)$ has an essential singularity suggests a third approach to calculating $f_1(x)$. We note that in the power-series expansion of $f_1(x)$

$$f_1(x) = \sum_{j=0}^{\infty} a_j x^j \tag{5.2}$$

only the coefficients a_j for large j are relevant for

calculating f_1 near the point where it has a singularity. Before discussing our conjecture for calculating a_j for large j, we will decompose $\bar{\Pi}_R(k)$ into a sum of terms $\Pi_n(k, \alpha)$ having the n electron-positron pair threshold. This decomposition will be useful for our discussion of the experimental implications of alternatives (i) and (ii) of Sec. I as well as for our treatment of the coefficient a_j for large j.

We thus write

$$\bar{\Pi}_R(k) = \sum_{n=0}^{\infty} \Pi_n(k, \alpha), \tag{5.3}$$

where $\mathrm{Im}\Pi_n(k, \alpha)$ includes all terms in the sum

$$\mathrm{Im}\bar{\Pi}_{\mu\nu}(k) \sim \sum_N \langle 0|j_\mu(0)|N\rangle\langle N|j_\nu(0)|0\rangle, \tag{5.4}$$

for which the state $|N\rangle$ contains n electron-positron pairs and any number of photons. Clearly

$$\mathrm{Im}\Pi_n(k, \alpha) \underset{k^2<4m^2n^2}{=} 0, \tag{5.5}$$

while

$$\Pi_n(k, \alpha) \underset{k^2 \gg 4m^2n^2}{\approx} \Pi_n^0(\alpha) + \Pi_n^1(\alpha)\ln\frac{k^2}{m^2} + \Pi_n^2(\alpha)\left(\ln\frac{k^2}{m^2}\right)^2 + \cdots. \tag{5.6}$$

Equations (5.3) and (5.6) combine to yield a reordered version of the asymptotic expression (4.5) for $\bar{\Pi}_R(k)$. Equation (5.6) is more precise than (4.5) in that it makes explicit the fact that as n increases one must go to larger and larger values of k in order to be in the asymptotic region.

The contribution to $\Pi_n(k, \alpha)$ which is of lowest order in α comes from the terms in the sum (5.4) for which the state $|N\rangle$ contains n electron positron pairs and no photons. The lowest-order perturbation value of $\langle 0|j^\mu|N\rangle$ for these states is proportional to α^{n-1}. This gives a contribution to $\Pi_n(k, \alpha)$ proportional to α^{2n-2}. For example, the lowest-order contribution to $\Pi_3(k, \alpha)$ comes from tree-graph contributions to $\langle 0|j_\mu|3e, 3p\rangle$ like those depicted in Fig. 13. Higher-order contributions to $\Pi_3(k, \alpha)$ come from radiative corrections to

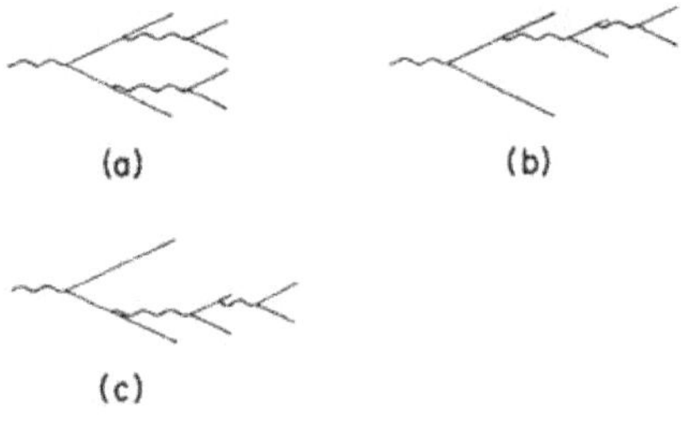

FIG. 13. Tree graphs for $\langle 0|j_\mu|3e, 3p\rangle$.

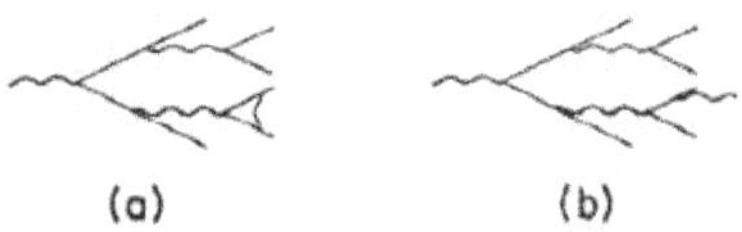

FIG. 14. Graphs for real and virtual radiative corrections to the graph of Fig. 13(a).

$\langle 0|j_\mu|3e, 3p\rangle$ such as that depicted in Fig. 14(a) as well as from lowest-order matrix elements $\langle 0|j_\mu|3e, 3p, 1 \text{ photon}\rangle$, depicted in Fig. 14(b). The amplitude of Fig. 14(a) gives a contribution to $\Pi_3(k, \alpha)$ containing infrared divergences which cancel the infrared divergences arising from the contribution to $\Pi_3(k, \alpha)$ of the amplitude of Fig. 14(b).

The single-closed-fermion-loop contribution to $\Pi_n(k, \alpha)$ for $k^2 \gg (2mn)^2$ behaves like a single power of $\ln(k^2/m^2)$ with a coefficient $\alpha^{2n-2}b_n(\alpha)$ which is the n-pair-state contribution to $f_1(\alpha)$. To calculate $b_n(\alpha)$ and hence $f_1(\alpha)$, we must evaluate the single-closed-loop contribution to the sum (5.4) with m set equal to zero.[19] Clearly states $|N\rangle$ which contain only photons do not contain any single-closed-loop contribution, i.e., $b_0(\alpha) = 0$. Furthermore the lowest-order n-pair contribution $\alpha^{2n-2}b_n(0)$ to $f_1(\alpha)$ arises from tree diagrams such as those depicted in Fig. 13. We can thus write $f_1(x)$ as a sum over n-pair contributions:

$$f_1(x) = \sum_{n=1}^{\infty} x^{2n-2}b_n(x). \tag{5.7}$$

Comparison of (5.7) with the power-series expansion (5.2) shows that the n-pair state contributes to all terms in the series (5.2) for which $j \geq 2n-2$. That is, every time j increases by 2, the coefficient a_j is increased by the contribution from a state with one additional pair.

We now conjecture that the dominant contribution to the coefficients a_j for large $j = 2n-2$ arises from the n-pair state in the sum (5.7), that is, for large n we can write

$$a_{2n-2} \sim b_n(0). \tag{5.8}$$

Of course a_{2n-2} also receives contributions from radiative corrections from states having fewer than n pairs via the diagrams depicted in Fig. 14. The assumption is that these radiative corrections do not contribute substantially to the growth of a_j for large j. We know that the contribution of these corrections in the infrared region cancels, and we see no physical reason why they should be important for large j. Thus our assumption (5.8) is that the origin of the essential singularity in $f_1(x)$ is the rapid growth of the lowest-order n-pair contribution $b_n(0)$ as n, the number of pairs, increases. That is, the essential singularity in $f_1(x)$ is due to the presence of the infinite number of multiparticle thresholds which is characteristic of a relativistic theory.

The above paragraph is of course pure speculation and is an attempt to give physical motivation for calculating the coefficients $b_n(0)$ for large n. To calculate $b_n(0)$ we must calculate the tree-graph contribution to the matrix element $\langle 0|j_\mu|n \text{ pairs}\rangle$, project the single-closed-loop contribution out of the product $\langle 0|j_\mu|n \text{ pairs}\rangle \times\langle n \text{ pairs}|j_\mu|0\rangle$, and finally integrate over $2n$-particle phase space. In carrying out this procedure, we, of course, use mass-zero electrons and positrons. We have shown that states in which each electron-positron pair is produced with zero total momentum contributes nothing to $b_n(0)$. The calculation of the contribution to $b_n(0)$ from more complicated kinematic configurations of n electron-positron pairs may be feasible, but we have not progressed very far with this calculation.

This concludes a summary of our current meager theoretical knowledge of $f_1(x)$ and we now speculate briefly on the implications of this discussion for high-energy experiments.

The question we pose in confronting alternatives (i) and (ii) of Sec. I with at least conceivable future experiment is the following: The renormalized perturbation expansion for $\bar{\Pi}_R$ contains terms which behave asymptotically like $\alpha \ln(k^2/m^2)$. Such terms modify the photon propagator in a way that can be experimentally detected in sufficiently accurate high-energy electron-electron or electron-positron scattering or electron-positron annihilation. On the other hand we know that if quantum electrodynamics is a finite theory via either mechanism (i) or (ii), then $\bar{\Pi}_R(k)\to$ constant, i.e., $\alpha\bar{D} \sim\alpha_0/k^2$ as $k^2\to\infty$. Hence, in such a case the $\alpha \ln(k^2/m^2)$ terms must not be present for sufficiently high energy. We then ask at what energy these logarithmic terms disappear. Our answer is the following: If the physically correct solution corresponds to alternative (i) with $\alpha_0^{-1}=x_0^{-1}\ll\alpha^{-1} \sim 137$ (e.g., if $\alpha_0=\frac{1}{4}$), then $\alpha\bar{D}$ will not approach its asymptotic limit α_0/k^2 until we reach experimentally unattainable superhigh energies of the order me^{137}. Thus we should expect no deviations from the renormalized perturbation-theory predictions due to electrodynamic effects in any foreseeable experiment. However, in alternative (ii), $\alpha=x_0$, we cannot rule out the possibility that $\alpha\bar{D}$ attains its asymptotic limit at an energy which may be experimentally accessible. Very roughly we might expect this energy to be of the order of magnitude mn_0 where states with n_0 electron-positron pairs give important contributions to $f_1(\alpha)$. (Perhaps $n_0\sim 137$.) However, in lieu of a calculation of f_1, we have no indication of what this energy is. However, if perturbation-theory logarithms disappear in accurate high-energy experiments, then this might be interpreted as information in favor of alternative (ii) and might at the same time give us some information about $f_1(\alpha)$.

The basic reason for the above distinction between (i) and (ii) is that the mechanism by which $\bar{\Pi}_R(k)$ is made finite as $k^2\to\infty$ is completely different in the two cases. In alternative (ii) the asymptotic expansion (4.5) gives a finite result for $\bar{\Pi}_R(k)$ as $k^2\to\infty$ because the coefficients $c_j(\alpha)$ of $[\ln(k^2/m^2)]^j$ all vanish when $\alpha=x_0$. In alternative (i) no conclusion can be drawn directly from (4.5) because one must sum the infinite set of diagrams corresponding to vacuum-polarization insertions in internal photon lines before taking the limit $k^2\to\infty$. When this is done, the different powers of $\ln(k^2/m^2)$ in Eq. (4.5) combine to form a new expression for Π in terms of α_0, which, for $\alpha_0=x_0$, is asymptotically finite. In this case the simplest way to find the rate at which $\bar{\Pi}_R(k^2)$ approaches its limiting finite value is to use the Gell-Mann–Low equation, which we will now briefly describe.

Making certain plausible assumptions about the behavior of perturbation theory integrals for the photon propagator when $m=0$,[4] Gell-Mann and Low[1] derived the following equation for the asymptotic behavior of the photon propagator $\bar{D}(k^2)$:

$$\ln\frac{k^2}{m^2}=\int_{q(\alpha)}^{k^2\alpha\bar{D}(k^2)}\frac{dx}{x^2\psi(x)}\,. \tag{5.9}$$

In Eq. (5.9), $q(\alpha)$ is the constant in the asymptotic expansion of $k^2\bar{D}(k^2)$. $q(\alpha)=\alpha+O(\alpha^2)$. The function $\psi(x)$ is a function which vanishes at root x_0 of the equation $f(x)=0$. In case (i) as $k^2\to\infty$, $k^2\alpha\bar{D}\to\alpha_0 =x_0$, and the integral in (5.9) diverges at the upper limit since $\psi(x_0)=0$. For small values of x, $\psi(x) =\frac{1}{2}\pi(\frac{2}{3}+x/2\pi+\cdots)$. For concreteness let us assume $x_0=\frac{1}{4}$. Let us then ask, "How large does k^2 have to become before $k^2\alpha\bar{D}$ reaches a value which is greater than $\frac{1}{8}$?" We rewrite Eq. (5.9) in the form

$$\ln\frac{k^2}{m^2}=\int_{q(\alpha)}^{1/8}\frac{dx}{x^2\psi(x)}+\int_{1/8}^{k^2\alpha\bar{D}}\frac{dx}{x^2\psi(x)}\,. \tag{5.10}$$

Since $\psi(x)$ is positive for $0<x<x_0=\frac{1}{4}$, we conclude from Eq. (5.10) that

$$\ln\frac{k^2}{m^2}\underset{k^2\alpha\bar{D}>1/8}{>}\int_{q(\alpha)}^{1/8}\frac{dx}{x^2\psi(x)}$$

$$\approx\frac{1}{\psi(0)}\left[\frac{1}{q(\alpha)}-8\right], \tag{5.11}$$

that is, if $\alpha_0=x_0=\frac{1}{4}$, the photon propagator will not come within a factor 2 of its asymptotic limit until

$$\frac{k^2}{m^2} \geq \exp\left[\frac{1}{\psi(0)}\left(\frac{1}{q(\alpha)} - 8\right)\right] \sim e^{137}. \qquad (5.12)$$

Thus we conclude that if $\alpha_0 = x_0$ is not very small, then the photon propagator does not start approaching its asymptotic limit until superhigh unattainable energies. Thus in alternative (i) we do not expect deviations from renormalized perturbation behavior at any experimentally attainable energy. Of course if x_0 is small, i.e., only slightly greater than α, one cannot arrive at the above conclusion. This is because the factor 8 on the right-hand side of Eq. (5.18) would have to be replaced by a factor of order $1/x_0 \sim 137$ and hence could not be neglected in comparison with that factor $1/q(\alpha) \sim 1/\alpha \sim 137$.

Now let us consider the alternative (ii), $\alpha = x_0$. In this case we can use directly the high-energy expansion (4.5) of perturbation theory. However, we must first make the decomposition (5.3) of $\bar{\Pi}_R(k)$, since as the thresholds become higher, one must go to higher values of k to reach the asymptotic region (5.6). If $\alpha = x_0$, then from (5.3), (5.6), and (4.5)

$$\sum_{n=1}^{\infty} \Pi_n^j(\alpha) = c_j(\alpha) = 0, \quad j = 1, 2, \ldots . \qquad (5.13)$$

(5.13) is the condition that the coefficient $c_j(\alpha)$ of $[\ln(k^2/m^2)]^j$ in the high-k expansion of $\bar{\Pi}_R(k)$ vanishes. The coefficients (5.13) vanish as a consequence of (a) the condition $f_1(\alpha) = 0$ and (b) the vanishing of the n-photon amplitudes in $m = 0$ electrodynamics. Let us rewrite Eq. (5.3) in the form

$$\bar{\Pi}_R(k) = \sum_{n=1}^{n_0} \Pi_n(k, \alpha) + \sum_{n=n_0+1}^{\infty} \Pi_n(k, \alpha), \qquad (5.14)$$

where n_0 is the number of pairs which is important in the expansion of $f_1(\alpha)$. Then for $k^2 \gg (2mn_0)^2$ we have, using (5.6),

$$\bar{\Pi}_R(k) \underset{k^2 \gg (2mn_0)^2}{\approx} \sum_{n=1}^{n_0}\left[\Pi_n^0(\alpha) + \Pi_n^1(\alpha)\ln\frac{k^2}{m^2} + \cdots\right] + \sum_{n=n_0+1}^{\infty} \Pi_n(k, \alpha). \qquad (5.15)$$

Then if we use Eq. (5.13), (5.15) becomes

$$\bar{\Pi}_R(k) \underset{k^2 \gg (2mn_0)^2}{\approx} \sum_{n=1}^{\infty} \Pi_n^0(\alpha) + \sum_{n=n_0+1}^{\infty}\left[\Pi_n(k, \alpha) - \Pi_n^0(\alpha) - \Pi_n^1(\alpha)\ln\frac{k^2}{m^2} \cdots\right]. \qquad (5.16)$$

The first summation on the right-hand side of Eq. (5.16) gives the asymptotic value $\bar{\Pi}_R(\infty)$. For each value of n the sum of the terms in the bracket under the second summation in Eq. (5.16) vanishes as $k^2 \to \infty$. Denoting this sum by $s_n(k, \alpha)$, we can write (5.16) as

$$\bar{\Pi}(\alpha) \underset{k^2 \gg (2mn_0)^2}{\approx} \bar{\Pi}_R(\infty) + \sum_{n=n_0+1}^{\infty} s_n(k, \alpha), \qquad (5.17)$$

where $s_n(k, \alpha) \to 0$ for k fixed as $n \to \infty$ and for n fixed as $k \to \infty$. $s_n(k, \alpha)$ is the n-pair contribution to the asymptotically vanishing part of the vacuum polarization. If we assume that the dominant contribution to $s_n(k, \alpha)$ comes from the same pair states that are important for $f_1(\alpha)$, we conclude that

$$\sum_{n=n_0+1}^{\infty} s_n(k, \alpha) < \epsilon \quad \text{independent of } k^2, \qquad (5.18)$$

where ϵ is of the order of magnitude of $\sum_{n=n_0+1}^{\infty} \alpha^{2n-2} b_n(\alpha)$, which is small by our choice of n_0. In writing (5.18) we have assumed that the convergence is uniform, i.e., ϵ does not depend upon k^2.

Thus if we carry out an experiment at an energy which probes values of $k^2 > (2mn_0)^2$, the logarithmic behavior of perturbation theory disappears and

$$\Pi_R(k^2, \alpha) \underset{k^2 \gg (2mn_0)^2}{\approx} \bar{\Pi}_R(k^2, \alpha) + \epsilon . \qquad (5.19)$$

As of now we have no way of estimating n_0. It also may be that the value of n_0 that enters into Eq. (5.19) is not the value of n_0 that is important for $f_1(\alpha)$.

In any case it is not impossible that the transition from the perturbation-theory behavior (4.5) to the behavior (5.19) could show up at experimentally accessible energies. If in fact the behavior (5.19) were observed, we could rule out alternative (i) with α_0 not close to α, and it would be likely that alternative (ii) is the physical solution of quantum electrodynamics.

---

*Work supported in part by the U.S. Atomic Energy Commission.

†John Simon Guggenheim Fellow 1972–1973.

‡Permanent address.

[1] M. Gell-Mann and F. E. Low, Phys. Rev. 95, 1300 (1954).

[2] K. Johnson, M. Baker, and R. Willey, Phys. Rev. 136, B1111 (1964).

[3]K. Johnson, R. Willey, and M. Baker, Phys. Rev. 163, 1699 (1967).

[4]M. Baker and K. Johnson, Phys. Rev. 183, 1292 (1969).

[5]M. Baker and K. Johnson, Phys. Rev. D 3, 2516 (1971).

[6]M. Baker and K. Johnson, Phys. Rev. D 3, 2541 (1971).

[7]Stephen L. Adler, Phys. Rev. D 5, 3021 (1972).

[8]In the remainder of this paper we will usually call x simply the coupling constant. x is normalized to include the same factors of π as the fine-structure constant.

[9]Everywhere in this paper the photon mass is set equal to zero from the start. The usual photon-mass infrared divergence arises from the second subtraction in the equation for the renormalized electron propagator. This subtraction never has to be carried out if one works in the gauge in which Z_2 is ultraviolet-finite.

[10]Since π is gauge-invariant, in any other gauge the divergences arising from the vertex insertions cancel those arising from the electron self-energy insertions which must be included if Π is calculated in a gauge in which S is not finite.

[11]For $k = 0$ the integral (2.7) diverges at the lower limit. For $k \neq 0$ there is no such divergence since $f(p^2/k^2) \to 0$ as $p^2 \to 0$. This is obvious since the Feynman integrals for Π contain a factor $1/(p+k)^2$ rather than the $1/p^2$ factor which was introduced for convenience in Eq. (2.7).

[12]Setting $x = x_0$ in Eq. (2.7) clearly also eliminates the divergence in $\Pi(0)$ for small p mentioned in Ref. 10.

[13]P. G. Federbush and K. Johnson, Phys. Rev. 120, 1296 (1960).

[14]To obtain all the single-loop diagrams we should also include the self-energy corrections to the electron and positron on the outer loop. Such corrections yield diagrams which are equivalent to those depicted in Fig. 12. The sum of Figs. 11(a) and 12(a) and 12(b) then generate the lowest-order contribution to photon-photon scattering. In the gauge in which Z_2 is finite, we can omit the diagrams of Figs. 12(a) and 12(b).

[15]If the exact value of ϵ turned out to be close to $-\frac{1}{2}$, then there could be an electrodynamic enhancement of a small, weak-interaction-induced electron-muon mass difference. That is, since as $p^2 \to \infty$ electrodynamics behaves like a free-field theory, at some distance Λ the weak interaction becomes the predominant interaction of the electron and muon. Let us assume that the effect of the weak interaction is to cut off the integrals of electrodynamics when the momentum p becomes greater than Λ^{-1}. In this case the cutoff Λ in Eq. (4.13) is a physical cutoff produced by the weak interactions. If we let m_{0e} and $m_{0\mu}$ be the masses of the electron and muon in the absence of electromagnetism, then from Eq. (4.13) m_{0e} and $m_{0\mu}$ are given in terms of the physical masses m_e and m_μ of the electron and muon by the expressions

$$m_{0e} = a m_e \left[\frac{m_e^2}{\Lambda^2}\right]^\epsilon ,$$

$$m_0 = a m_\mu \left[\frac{m_\mu^2}{\Lambda^2}\right]^\epsilon ,$$

or

$$\frac{m_\mu}{m_e} = \left(\frac{m_{0\mu}}{m_{0e}}\right)^{1/(1+2\epsilon)}$$

Thus a small deviation from unity of $m_{0\mu}/m_{0e}$ produced by the weak interactions could be enhanced by electrodynamics into the large observed ratio m_μ/m_e if ϵ is close to $-\frac{1}{2}$.

[16]J. L. Rosner, Phys. Rev. Lett. 17, 1190 (1966).

[17]Howard Edward Brandt, Ph.D. thesis, University of Washington, 1970 (unpublished).

[18]Refat Abdellatif, Ph.D. thesis, University of Washington, 1970 (unpublished).

[19]Because we are evaluating only the single-closed-loop part of Eq. (5.4), this formula imposes no positivity requirement on $f_1(\alpha)$.

REFERENCES

Bjorken, J. D., Drell, S. D., 1964, *Relativistic Quantum Mechanics* (McGraw-Hill, New York, 1965).

Bjorken, J. D., Drell, S. D., 1965, *Relativistic Quantum Fields* (McGraw-Hill, New York, 1965).

Blaha, S., 1998, *Cosmos and Consciousness* (Pingree-Hill Publishing, Auburn, NH, 1998).

_______, 2002, *A Finite Unified Quantum Field Theory of the Elementary Particle Standard Model and Quantum Gravity Based on New Quantum Dimensions™ & a New Paradigm in the Calculus of Variations* (Pingree-Hill Publishing, Auburn, NH, 2002).

_______, 2003, *A Finite Unified Quantum Field Theory of the Elementary Particle Standard Model and Quantum Gravity Based on New Quantum Dimensions™ and a New Paradigm in the Calculus of Variations* (Pingree-Hill Publishing, Auburn, NH, 2003).

_______, 2004, *Quantum Big Bang Cosmology: Complex Space-time General Relativity, Quantum Coordinates™ Dodecahedral Universe, Inflation, and New Spin 0, ½, 1 & 2 Tachyons & Imagyons* (Pingree-Hill Publishing, Auburn, NH, 2004).

______, *2005a, Quantum Theory of the Third Kind: A New Type of Divergence-free Quantum Field Theory Supporting a Unified Standard Model of Elementary Particles and Quantum Gravity based on a New Method in the Calculus of Variations* (Pingree-Hill Publishing, Auburn, NH, 2005).

______, 2005b, *The Metatheory of Physics Theories, and the Theory of Everything as a Quantum Computer Language* (Pingree-Hill Publishing, Auburn, NH, 2005).

______, 2005c, *The Equivalence of Elementary Particle Theories and Computer Languages: Quantum Computers, Turing Machines, Standard Model, Superstring Theory, and a Proof that*

Gödel's Theorem Implies Nature Must Be Quantum (Pingree-Hill Publishing, Auburn, NH, 2005).

______, 2006a, *The Foundation of the Forces of Nature* (Pingree-Hill Publishing, Auburn, NH, 2006).

______, 2006b, *A Derivation of ElectroWeak Theory based on an Extension of Special Relativity; Black Hole Tachyons; & Tachyons of Any Spin.* (Pingree-Hill Publishing, Auburn, NH, 2006).

_______, 2007a, *Physics Beyond the Light Barrier: The Source of Parity Violation, Tachyons, and A Derivation of Standard Model Features* (Pingree-Hill Publishing, Auburn, NH, 2007).

______, 2007b, *The Origin of the Standard Model: The Genesis of Four Quark and Lepton Species, Parity Violation, the ElectroWeak Sector, Color SU(3), Three Visible Generations of Fermions, and One Generation of Dark Matter with Dark Energy* (Pingree-Hill Publishing, Auburn, NH, 2007).

______*, 2008a, A Direct Derivation of the Form of the Standard Model From GL(16) (Pingree-Hill Publishing, Auburn, NH, 2008).*

______, 2008b, *A Complete Derivation of the Form of the Standard Model With a New Method to Generate Particle Masses Second Edition* (Pingree-Hill Publishing, Auburn, NH, 2008)

______, 2009, *The Algebra of Thought & Reality: The Mathematical Basis for Plato's Theory of Ideas, and Reality Extended to Include A Priori Observers and Space-Time Second Edition* (Pingree-Hill Publishing, Auburn, NH, 2009).

______, 2010a, *Operator Metaphysics: A New Metaphysics Based on a New Operator Logic and a New Quantum Operator Logic that Lead to a Mathematical Basis for Plato's Theory of Ideas and Reality* (Pingree-Hill Publishing, Auburn, NH, 2010).

______, 2010b, *The Standard Model's Form Derived from Operator Logic, Superluminal Transformations and GL(16)* (Pingree-Hill Publishing, Auburn, NH, 2010).

______, 2010c, *SuperCivilizations: Civilizations as Superorganisms* (McMann-Fisher Publishing, Auburn, NH, 2010).

______, 2011a, *21st Century Natural Philosophy Of Ultimate Physical Reality* (McMann-Fisher Publishing, Auburn, NH, 2011).

______, 2011b, *All the Universe! Faster Than Light Tachyon Quark Starships & Particle Accelerators with the LHC as a Prototype Starship Drive Scientific Edition* (Pingree-Hill Publishing, Auburn, NH, 2011).

______, 2011c, *From Asynchronous Logic to The Standard Model to Superflight to the Stars* (Blaha Research, Auburn, NH, 2011).

______, 2012a, *From Asynchronous Logic to The Standard Model to Superflight to the Stars volume 2: Superluminal CP and CPT, U(4) Complex General Relativity and The Standard Model, Complex Vierbein General Relativity, Kinetic Theory, Thermodynamics* (Blaha Research, Auburn, NH, 2012).

______, 2012b, *Standard Model Symmetries, And Four And Sixteen Dimension Complex Relativity; The Origin Of Higgs Mass Terms* (Blaha Reasearch, Auburn, NH, 2012).

______, 2013a, *Multi-Stage Space Guns, Micro-Pulse Nuclear Rockets, and Faster-Than-Light Quark-Gluon Ion Drive Starships* (Blaha Research, Auburn, NH, 2013).

______, 2013b, *The Bridge to Dark Matter; A New Sister Universe; Dark Energy; Inflatons; Quantum Big Bang; Superluminal Physics; An Extended Standard Model Based on Geometry* (Blaha Reasearch, Auburn, NH, 2013).

______, 2014a, *Universes and Megaverses: From a New Standard Model to a Physical Megaverse; The Big Bang; Our Sister Universe's Wormhole; Origin of the Cosmological Constant, Spatial Asymmetry of the Universe, and its Web of Galaxies; A Baryonic Field between Universes and Particles; Megaverse Extended Wheeler-DeWitt Equation* (Blaha Reasearch, Auburn, NH, 2014).

______, 2014b, *All the Megaverse! Starships Exploring the Endless Universes of the Cosmos Using the Baryonic Force* (Blaha Research, Auburn, NH, 2014).

______, 2014c, *All the Megaverse! II Between Megaverse Universes: Quantum Entanglement Explained by the Megaverse Coherent Baryonic Radiation Devices – PHASERs Neutron Star Megaverse Slingshot Dynamics Spiritual and UFO Events, and the Megaverse Microscopic Entry into the Megaverse* (Blaha Research, Auburn, NH, 2014).

______, 2015a, *PHYSICS IS LOGIC PAINTED ON THE VOID: Origin of Bare Masses and The Standard Model in Logic, U(4) Origin of the Generations, Normal and Dark Baryonic Forces, Dark Matter, Dark Energy, The Big Bang, Complex General Relativity, A Megaverse of Universe Particles* (Blaha Research, Auburn, NH, 2015).

______, 2015b, *PHYSICS IS LOGIC Part II: The Theory of Everything, The Megaverse Theory of Everything, U(4)⊗U(4) Grand Unified Theory (GUT), Inertial Mass = Gravitational Mass, Unified Extended Standard Model and a New Complex General Relativity with Higgs Particles, Generation Group Higgs Particles* (Blaha Research, Auburn, NH, 2015).

______, 2015c, *The Origin of Higgs ("God") Particles and the Higgs Mechanism: Physics is Logic III, Beyond Higgs – A Revamped Theory With a Local Arrow of Time, The Theory of Everything Enhanced, Why Inertial Frames are Special, Universes of the Mind* (Blaha Research, Auburn, NH, 2015).

______, 2015d, *The Origin of the Eight Coupling Constants of The Theory of Everything: U(8) Grand Unified Theory of Everything (GUTE), S^8 Coupling Constant Symmetry, Space-Time Dependent Coupling Constants, Big Bang Vacuum Coupling Constants, Physics is Logic IV* (Blaha Research, Auburn, NH, 2015).

______, 2016a, *New Types of Dark Matter, Big Bang Equipartition, and A New U(4) Symmetry in the Theory of Everything: Equipartition Principle for Fermions, Matter is 83.33% Dark, Penetrating the Veil of the Big Bang, Explicit QFT Quark Confinement and Charmonium, Physics is Logic V* (Blaha Research, Auburn, NH, 2016).

______, 2016b, *The Periodic Table of the 192 Quarks and Leptons in The Theory of Everything: The U(4) Layer Group, Physics is Logic VI* (Blaha Research, Auburn, NH, 2016).

______, 2016c, *New Boson Quantum Field Theory, Dark Matter Dynamics, Dark Matter Fermion Layer Mixing, Genesis of Higgs Particles, New Layer Higgs Masses, Higgs Coupling Constants, Non-Abelian Higgs Gauge Fields, Physics is Logic VII* (Blaha Research, Auburn, NH, 2016).

______, 2016d, *Unification of the Strong Interactions and Gravitation: Quark Confinement Linked to Modified Short-Distance Gravity; Physics is Logic VIII* (Blaha Research, Auburn, NH, 2016).

______, 2016e, *MoND: Unification of the Strong Interactions and Gravitation II, Quark Confinement Linked to Large-Scale Gravity, Physics is Logic IX* (Blaha Research, Auburn, NH, 2016).

______, 2016f, *CQ Mechanics: A Unification of Quantum & Classical Mechanics, Quantum/Semi-Classical Entanglement, Quantum/Classical Path Integrals, Quantum/Classical Chaos* (Blaha Research, Auburn, NH, 2016).

______, 2016g, *GEMS: Unified Gravity, ElectroMagnetic and Strong Interactions: Manifest Quark Confinement, A Solution for the Proton Spin Puzzle, Modified Gravity on the Galactic Scale* (Pingree Hill Publishing, Auburn, NH, 2016).

______, 2016h, *Unification of the Seven Boson Interactions based on the Riemann-Christoffel Curvature Tensor* (Pingree Hill Publishing, Auburn, NH, 2016).

______, 2017a, *Unification of the Eleven Boson Interactions based on 'Rotations of Interactions'* (Pingree Hill Publishing, Auburn, NH, 2017).

______, 2017b, *The Origin of Fermions and Bosons, and Their Unification* (Pingree Hill Publishing, Auburn, NH, 2017).

______, 2017c, *Megaverse: The Universe of Universes* (Pingree Hill Publishing, Auburn, NH, 2017).

______, 2017d, *SuperSymmetry and the Unified SuperStandard Model* (Pingree Hill Publishing, Auburn, NH, 2017).

______, 2017e, *From Qubits to the Unified SuperStandard Model with Embedded SuperStrings: A Derivation* (Pingree Hill Publishing, Auburn, NH, 2017).

______, 2017f, *The Unified SuperStandard Model in Our Universe and the Megaverse: Quarks, ...* , (Pingree Hill Publishing, Auburn, NH, 2017).

______, 2018a, *The Unified SuperStandard Model and the Megaverse SECOND EDITION A Deeper Theory based on a New Particle Functional Space that Explicates Quantum Entanglement Spookiness (Volume 1)* (Pingree Hill Publishing, Auburn, NH, 2018).

______, 2018b, *Cosmos Creation: The Unified SuperStandard Model, Volume 2, SECOND EDITION* (Pingree Hill Publishing, Auburn, NH, 2018).

______, 2018c, *God Theory (*Pingree Hill Publishing, Auburn, NH, 2018).

______, 2018d, *Immortal Eye: God Theory: Second Edition* (Pingree Hill Publishing, Auburn, NH, 2018).

______, 2018e, *Unification of God Theory and Unified SuperStandard Model THIRD EDITION* (Pingree Hill Publishing, Auburn, NH, 2018).

______, 2019, *Calculation of: QED α = 1/137, and Other Coupling Constants of the Unified SuperStandard Theory* (Pingree Hill Publishing, Auburn, NH, 2019).

Gradshteyn, I. S. and Ryzhik, I. M., 1965, *Table of Integrals, Series, and Products* (Academic Press, New York, 1965).

Heitler, W., 1954, *The Quantum Theory of Radiation* (Claendon Press, Oxford, UK, 1954).

Huang, Kerson, 1992, *Quarks, Leptons & Gauge Fields 2nd Edition* (World Scientific Publishing Company, Singapore, 1992).

INDEX

About the Author

Stephen Blaha is a well known Physicist and Man of Letters with interests in Science, Society and civilization, the Arts, and Technology. He had an Alfred P. Sloan Foundation scholarship in college. He received his Ph.D. in Physics from Rockefeller University. He has served on the faculties of several major universities. He was also a Member of the Technical Staff at Bell Laboratories, a manager at the Boston Globe Newspaper, a Director at Wang Laboratories, and President of Blaha Software Inc and of Janus Associates Inc. (NH).

Among other achievements he was a co-discoverer of the "r potential" for heavy quark binding developing the first (and still the only demonstrable) non-abelian gauge theory with an "r" potential; first suggested the existence of topological structures in superfluid He-3; first proposed Yang-Mills theories would appear in condensed matter phenomena with non-scalar order parameters; first developed a grammar-based formalism for quantum computers and applied it to elementary particle theories; first developed a new form of quantum field theory without divergences (thus solving a major 60 year old

problem that enabled a unified theory of the Standard Model and Quantum Gravity without divergences to be developed); first developed a formulation of complex General Relativity based on analytic continuation from real space-time; first developed a generalized non-homogeneous Robertson-Walker metric that enabled a quantum theory of the Big Bang to be developed without singularities at t = 0; first generalized Cauchy's theorem and Gauss' theorem to complex, curved multi-dimensional spaces; received Honorable Mention in the Gravity Research Foundation Essay Competition in 1978; first developed a physically acceptable theory of faster-than-light particles; first derived a composition of extrema method in the Calculus of Variations; first quantitatively suggested that inflationary periods in the history of the universe were not needed; first proved Gödel's Theorem implies Nature must be quantum; provided a new alternative to the Higgs Mechanism, and Higgs particles, to generate masses; first showed how to resolve logical paradoxes including Gödel's Undecidability Theorem by developing Operator Logic and Quantum Operator Logic; first developed a quantitative harmonic oscillator-like model of the life cycle, and interactions, of civilizations; first showed how equations describing superorganisms also apply to civilizations. A recent book shows his theory applies successfully to the past 14 years of history and to *new* archaeological data on Andean and Mayan civilizations as well as Early Anatolian and Egyptian civilizations.

He first developed an axiomatic derivation of the form of The Standard Model from geometry – space-time properties – The Unified SuperStandard Model. It unifies all the known forces of Nature. It also has a Dark Matter sector that includes a Dark ElectroWeak sector with Dark doublets and Dark gauge interactions. It uses quantum coordinates to remove infinities that crop up in most interacting quantum field theories and additionally to remove the infinities that appear in the Big Bang and generate inflationary growth of the universe. It shows gravity has a MOND-like form without sacrificing Newton's Laws. It relates the interactions of the MOND-like sector of gravity with the r-potential of Quark Confinement. The axioms of the theory lead to the question of their origin. We suggest in the preceding edition of this book it can be attributed to an entity with God-like properties. We explore these properties in "God Theory" and show they predict that the Cosmos exists forever although individual universes (or incarnations of our universe) "come and go." Several other important results emerge from God Theory such a a functionally triune God. The Unified SuperStandard Theory has many other important parts described in the Current Edition of *The Unified Superstandard Theory* and expanded in subsequent volumes.

Blaha has had a major impact on a succession of elementary particle theories: his Ph.D. thesis (1970), and papers, showed that quantum field theory calculations to all orders in ladder approximations could not give scaling deep inelastic electron-nucleon scattering. He later showed the eigenvalue equation for the fine structure constant α in Johnson-Baker-Willey QED had a zero at $\alpha = 1$ not 1/137 by solving the Schwinger-Dyson equations to all orders in an approximation that agreed with exact results to 4th

order in α thus ending interest in this theory. In 1979 at Prof. Ken Johnson's (MIT) suggestion he calculated the proton-neutron mass difference in the MIT bag model and found the result had the wrong sign reducing interest in the bag model. These results all appear in Physical Review papers. In the 2000's he repeatedly pointed out the shortcomings of SuperString theory and showed that The Standard Model's form could be derived from space-time geometry by an extension of Lorentz transformations to faster than light transformations. This deeper space-time basis greatly increases the possibility that it is part of THE fundamental theory.Recently, Blaha showed that the Weak interactions differed significantly from the Strong, electromagnetic and gravitation interactions in important respects while these interactions had similar features, and suggested that ElectroWeak theory, which is essentially a glued union of the Weak interactions and Electromagnetism, possibly modulo unknown Higgs particle features, be replaced by a unified theory of the other interactions combined with a stand-alone Weak interaction theory. Blaha also showed that, if Charmonium calculations are taken seriously, the Strong interaction coupling constant is only a factor of five larger than the electromagnetic coupling constant, and thus Strong interaction perturbation theory would make sense and yield physically meaningful results.

In graduate school (1965-71) he wrote substantial papers in elementary particles and group theory: The Inelastic E- P Structure Functions in a Gluon Model. Phys. Lett. B40:501-502,1972; Deep-Inelastic E-P Structure Functions In A Ladder Model With Spin 1/2 Nucleons, Phys.Rev. D3:510-523,1971; Continuum Contributions To The Pion Radius, Phys. Rev. 178:2167-2169,1969; Character Analysis of U(N) and SU(N), J. Math. Phys. 10, 2156 (1969); and The Calculation of the Irreducible Characters of the Symmetric Group in Terms of the Compound Characters, (Published as Blaha's Lemma in D. E. Knuth's book: *The Art of Computer Programming Vols. 1 – 4*).

In the early 1980's Blaha was also a pioneer in the development of UNIX for financial, scientific and Internet applications: benchmarked UNIX versions showing that block size was critical for UNIX performance, developing financial modeling software, starting database benchmarking comparison studies, developing Internet-like UNIX networking (1982) and developing a hybrid shell programming technique (1982) that was a precursor to the PERL programming language. He was also the manager of the AT&T ten-year future products development database. His work helped lead to commercial UNIX on computers such as Sun Micros, IBM AIX minis, and Apple computers.

In the 1980's he pioneered the development of PC Desktop Publishing on laser printers. and was nominated for three "Awards for Technical Excellence" in 1987 by PC Magazine for PC software products that he designed and developed.

Recently he has developed a theory of Megaverses – actual universes of which our universe is one – with quantum particle-like properties based on the Wheeler-DeWitt equation of Quantum Gravity. He has developed a theory of a baryonic force, which had been conjectured many years ago, and estimated the strength of the force based on discrepancies in measurements of the gravitational constant G. This force, operative in D-dimensinal space, can be used to escape from our universe in "uniships" which are the equivalent of the faster-than-light starships proposed in the author's earlier books. Thus travel to other universes, as well as to other stars is possible.

Blaha also considered the complexified Wheeler-DeWitt equation and showed that its limitation to real-valued coordinates and metrics generated a Cosmological Constant in the Einstein equations.

The author has also recently written a series of books on the serious problems of the United States and their solution as well as a book on the decline of Mankind that will follow from current social and genetic trends in Mankind.

In the past twelve years Dr. Blaha has written over 40 books on a wide range of topics. Some recent major works are: *From Asynchronous Logic to The Standard Model to Superflight to the Stars*, *All the Universe!*, *SuperCivilizations: Civilizations as Superorganisms*, *America's Future: an Islamic Surge, ISIS, al Qaeda, World Epidemics, Ukraine, Russia-China Pact, US Leadership Crisis*, *The Rises and Falls of Man – Destiny – 3000 AD: New Support for a Superorganism MACRO-THEORY of CIVILIZATIONS From CURRENT WORLD TRENDS and NEW Peruvian, Pre-Mayan, Mayan, Anatolian, and Early Egyptian Data, with a Projection to 3000 AD*, and *Mankind in Decline: Genetic Disasters, Human-Animal Hybrids, Overpopulation, Pollution, Global Warming, Food and Water Shortages, Desertification, Poverty, Rising Violence, Genocide, Epidemics, Wars, Leadership Failure.*

He has taught approximately 4,000 students in undergraduate, graduate, and postgraduate corporate education courses primarily in major universities, and large companies and government agencies.

www.ingramcontent.com/pod-product-compliance
Lightning Source LLC
Chambersburg PA
CBHW082007190326
41458CB00010B/3110
* 9 7 8 1 7 3 2 8 2 4 5 4 6 *